The Neuronomicon

Grayson W. Hooper
Editor

The Neuronomicon

A Guide to Acute Neuroradiological Pathology

Springer

Editor
Grayson W. Hooper
Department of Neuroradiology
UH/Case Western Reserve University School of Medicine
Cleveland, OH, USA

ISBN 978-3-032-05384-8 ISBN 978-3-032-05385-5 (eBook)
https://doi.org/10.1007/978-3-032-05385-5

This Springer imprint is published by the registered company Springer Nature Switzerland AG
The registered company address is: Gewerbestrasse 11, 6330 Cham, Switzerland

Preface

I wrote this book to help the radiologist-in-training and create a foundation for the neuroradiology fellow. The material herein is curated to represent critical cases that one will routinely experience while on service as a neuroradiology fellow and/or a radiology trainee on call. Most of these cases represent pathologies that one must recognize immediately and effectively communicate to your team members. Some are less urgent but represent material that may be valuable during clinical rotations or testable on major examinations.

The book contains images of multiple disease processes, many collected by Dr. Michael Coffey over a lifetime of radiology practice. Study the images and the figure legends. I selected them as "archetype" case presentations that hopefully help you identify similar findings while you're on service.

Each section follows a pattern of presentation with some deviation in the trauma chapter. The nitty-gritty may not be immediately useful to you while you're on call and under pressure for a read, or if an attending has you in their grasp during readout. However, I encourage you to read it to create a strong foundation. For example, being able to fluently discuss specific vascular territories, clinical presentations, and associated factoids of stroke not only helps build confidence in your skills in the referral community, but it may also prove useful during exams. If you plan to pursue neurointerventional radiology, you must familiarize yourself with the body of literature behind it, such as NASCET, DAWN, DEFUSE 3, EXTEND, and SELECT 2 to name a few—trust me, I found out the hard way the cost of ignorance. Hi, Dr. S!

Cleveland, OH, USA Grayson W. Hooper

About the Author

I am nobody. Who are you?

Acknowledgments

For:

God for giving me the opportunity to be.

My family for giving me the opportunity to be healthy.

Dickinson College for giving me the opportunity to be educated.

WVSOM Lewisburg for giving me the opportunity to be a physician.

The US Army and TAMC for giving me the opportunity to be a radiologist.

UH Cleveland/CWRU for giving me the opportunity to be a neuroradiologist.

Michael Coffey, MD: thank you for your case contributions and instruction, which were instrumental in the preparation of this book.

Travis Whilden: thank you for helping me create and refine the graphics critical to this book.

Wassim Malak, MD: thank you for reviewing Chap. 3 in great detail and providing feedback.

Tim Biega, MD; Ricardo Burgos, MD; Felix Song, MD: thank you for teaching me as an ornery radiology resident and introducing me to neuroradiology.

Ari Blitz, MD: thank you for consistent encouragement in this project as well as your teaching.

Anne Jay, MD: thank you for reviewing Chap. 1 and providing feedback.

Salvatore Labruzzo, DO: thank you for reviewing Chap. 1 and providing feedback.

Daniel Ginat, MD: thank you for involving me in academic projects over the course of my career.

Michael Lurakis, DO: thank you for being my earliest mentor while navigating the medical field.

Sven Dekeyzer, MD: thank you for your amazing online contributions to neuroradiology education and support of this textbook.

The crew of BlueSky Brews in Cleveland, OH: thank you for keeping this old soul well caffeinated and ethanol'd when I needed a Guinness.

Glory to Ukraine! **Слава Україні!**

Ethics Approval This book was produced in line with the principles of the Declaration of Helsinki. No human or animal research was conducted for the creation of this book. IRB exemption provided under STUDY20241381 by University Hospitals Cleveland Medical Center/Case Western Reserve University prior to publication. All images for this manuscript have been anonymized and undergone post hoc quality control to ensure patient privacy and adherence to US federal regulations (HIPAA).

Competing Interests None.

About This Book

The Neuronomicon was designed around the emergencies and urgencies of neuroradiology. There are both focused and general concepts within these pages that I believe a safe and competent radiologist should know. I tried to include as many pearls as possible, which may prove useful while on service or sitting for an exam.

If you have time, read everything. I have selected every word and image carefully based on feedback from people *waaaaaaaayyyy* smarter than me, as well as personal experience. Believe me, I have screwed up / missed / and misdiagnosed many disease processes during my youthful training years, but such is part of the process. However, if I can impart knowledge to others such that they can avoid certain pitfalls, landmines, and booby-traps that I myself have fell in, then the world of radiology would have become a better place and this book worthwhile.

If you are time constrained, then look at the images and read the figure legends. Establishing "templates" in your mind for normal and abnormal is essential for speed and accuracy in this field. As of 2025, there is no other way to obtain these skills aside from reading and frequent exposure.

I welcome all feedback regarding the images, presentation, and information contained within this book. Do you have a better idea on presentation format? Email me. Is something unclear? Email me. Do you see an error? Email me. If subsequent editions ever materialize, I'll include the changes and credit you.

Groovy

Contents

Contributors

Troy Holsworth WVSOM, Lewisburg, WV, USA

Grayson W. Hooper Department of Neuroradiology, UH/Case Western Reserve University School of Medicine, Cleveland, OH, USA

Shruti Kumari Department of Radiology, Case Western/UH, Cleveland, OH, USA

Abbreviations

ACA	Anterior Cerebral Artery
AchA	Anterior Choroidal Artery
Acomm	Anterior Communicating Artery
ADC	Apparent Diffusion Coefficient
ADEM	Acute Disseminated Encephalomyelitis
AHL	Acute Hemorrhagic Leukoencephalitis
AHT	Abusive Head Trauma
AICA	Anterior Inferior Cerebellar Artery
AoP	Artery of Percheron
AQP4	Aquaporin 4 Water Channel
ARIA	Amyloid Related Imaging Abnormality
AVM	Arteriovenous Malformation
BA	Brodmann Area
BAMF	Badass Motherfu****
CAA	Cerebral Amyloid Angiopathy
CBF	Cerebral Blood Flow
CBV	Cerebral Blood Volume
CCF	Cavernous Carotid Fistula
CIDP	Chronic Inflammatory Demyelinating Polyneuropathy
CJD	Creutzfeldt-Jakob disease
CLOCC	Cytotoxic Lesion of the Corpus Callosum
CM	Cryptococcal Meningitis
CMV	Cytomegalovirus
CN	Cranial Nerve
CNS	Central Nervous System
CO	Carbon Monoxide
CSF	Cerebrospinal Fluid
CT	Computed Tomography
CTA	CT Angiography
DAI	Diffuse Axonal Injury
DiS	Dissemination in Space
DiT	Dissemination in Time
DNL	Disseminated Necrotizing Leukoencephalopathy
DSA	Digital Subtraction Angiography

DTI	Diffusion Tensor Imaging
DWI	Diffusion Weighted Imaging
ECA	External Carotid Artery
EG	Ethylene Glycol
FLAIR	Fluid Attenuation Inversion Recovery
GBS	Guillain-Barré Syndrome
GCS	Glasgow Coma Score
HE	Hepatic Encephalopathy
HHV-6	Human Herpes Virus 6
HIV	Human Immunodeficiency Virus
HM	Hemiplegic Migraine
HPI	Hypoperfusion Index
HSV	Herpes Simplex Virus
HU	Hounsfield Unit
ICA	Internal Carotid Artery
IRIS	Immune Reconstitution Inflammatory Syndrome
JC	John Cunningham (virus)
LA	Ludwig's Angina
LGN	Lateral Geniculate Nucleus
LP	Lumbar Puncture
LVO	Large Vessel Occlusion
MCA	Middle Cerebral Artery
MeVO	Medium Vessel Occlusion
MIP	Maximum Intensity Projection
MLF	Medial Longitudinal Fasciculus
MOGAD	Myelin Oligodendrocyte Glyoprotein Antibody Disease
MRA	MR Angiography
mRS	Modified Rankin Scale
MS	Multiple Sclerosis
MTT	Mean Transit Time
MTX	Methotrexate
NAT	Nonaccidental Trauma
NBD	Neuro-Behçet's Disease
NIHSS	National Institutes of Health Stroke Scale
NMO-SD	Neuromyelitis Optica Spectrum Disorder
NOE	Naso-Orbito-Ethmoid
OAS	Orbital Apex Syndrome
OO	Os Odontoideum
ORIF	Open Reduction Internal Fixation
PACNS	Primary Angiitis of the CNS
PCA	Posterior Cerebral Artery
PCOM	Posterior Communicating Artery
PEP	Postexposure Prophylaxis
PICA	Posterior Inferior Cerebellar Artery
PLIC	Posterior Limb of the Internal Capsule

PML	Progressive Multifocal Leukoencephalopathy
PPRF	Pontine Paramedian Reticular Formation
PRES	Posterior Reversible Encephalopathy Syndrome
PTA	Peritonsillar Abscess
RA	Recurrent Artery (of Heubner)
RCVS	Reversible Cerebral Vasoconstriction Syndrome
RIL	Radiation Induced Leukoencephalopathy
SCA	Superior Cerebellar Artery
SCIWORA	Spinal Cord Injury Without Radiologic Abnormality
SIH	Spontaneous Intraparenchymal Hemorrhage
SMA	Supplementary Motor Area
SOFS	Superior Orbital Fissure Syndrome
SSPE	Subacute Sclerosing Panencephalitis
SSS	Superior Sagittal Sinus
SWI	Susceptibility Weighted Imaging
TBE	Tick-Borne Encephalitis
TBM	Tuberculous Meningoencephalitis
Tmax	Time to Maximum
TOF	Time-of-Flight
TSE	Transmissible Spongiform Encephalopathy
VAD	Vertebral Artery Dissection
VZO	Varicella Zoster Ophthalmicus
ZMC	Zygomaticomaxillary Complex

Stroke

1

Grayson W. Hooper

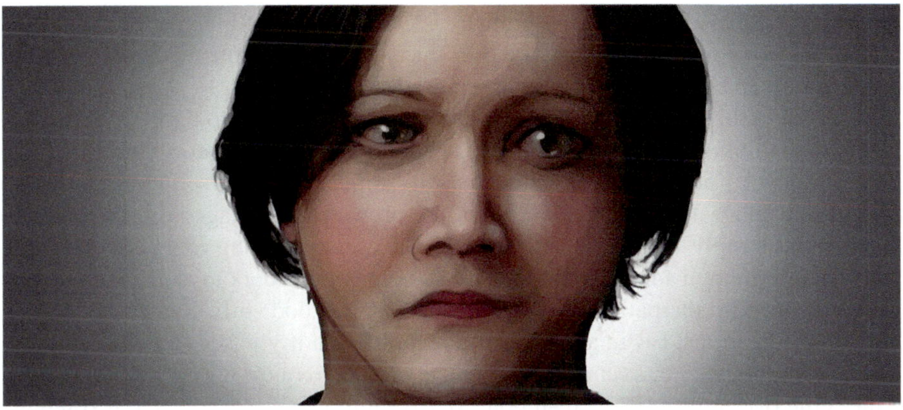

Stroke

Abstract

The chapter begins with the radiologic workup of ischemic stroke, including basic noncontrast head CT, CT angiography, and MR diffusion weighted imaging. It covers the basic, salient anatomy that a radiology resident should know with respect to the anterior and posterior circulations as well as the Circle of Willis. Further material delves into in-depth vascular, territorial, and functional anatomy, including Brodmann Areas, and stroke presentations. This

G. W. Hooper (✉)
Department of Neuroradiology, UH/Case Western Reserve University School of Medicine, Cleveland, OH, USA

G. W. Hooper (ed.), *The Neuronomicon*,
https://doi.org/10.1007/978-3-032-05385-5_1

information is presented in a clinical-radiologic context, describing unique syndromes, eponyms, and material that could also be useful while at the stat desk or on clinical rotations. The ischemic stroke section is concluded with a discussion of the DAWN, DEFUSE 3, and EXTEND trials and stroke perfusion imaging. Common pitfalls are discussed, such as stroke mimics and contrast staining.

Venous infarcts are then reviewed with close anatomic correlation provided courtesy of licensed Radiopaedia imaging from Dr. Craig Hacking. These include cortical vein and dural venous sinus thromboses. Diagnostic utility of CT and MR venography is reviewed, as are the pitfalls present in time-of-flight MRI. Finally, the chapter ends with a review of hemorrhagic stroke etiologies to include aneurysm, AVM, RCVS, amyloid angiopathy, PRES, and ARIA.

Stroke, cerebrovascular accident, brain attack, *apoplexia*—all words for the unfortunate phenomenon of brain destruction that has brought sorrow, disability, and death to human beings since time immemorial.

My goal in this section is not necessarily to teach you *what* to think. It's to teach you *how* to think. When you are on call, whether that be in a rural community hospital or a level 1 academic trauma center with robot concierge services, you must take into account available resources and be prepared to engage with the clinical team. Don't simply sit in your comfortable, dark, cozy reading room and interpret in a vacuum (though admittedly I do love the comfortable dark myself). Pick up the phone or open up the patient's chart and dig for the good stuff. Symptomatology, history, medications, ER intake notes—use it all! And if you see something out of place, be proactive. For example, if you receive a stat request for an MRI and know it takes the overnight MRI tech an hour to drive into work, alert the team. You may change management, which brings me to my list of **stroke call imperatives**!

Additionally, I recommend that you read this chapter while listening to *Tool*'s **Lateralus** album, which—in this humble author's opinion—is literally the best thing since penicillin.

Read the below imperatives in the voice of Immortan Joe.

Imperative 1 Time is brain. Don't dawdle with other studies, administrative tasks, or technical difficulties. Read the stroke studies immediately. If your PACS software crashes, go to the CT suite and read from their terminals.

Imperative 2 Do not obstruct. Give the team what they ask for. And if something is markedly out of the ordinary or dangerous, call your staff and seek guidance, even if it is 3:00 AM on Christmas morning (true story).

Imperative 3 Do your duty. Stroke is a clinical diagnosis and noncontrast head CTs serve the purpose of excluding contraindications to chemical thrombolysis—bleeding and mass. Sure, look for loss of gray-white differentiation; scrutinize the

Sylvian fissures for dense vessels; and search the tentorial leaflets for tiny hemorrhages. Just don't do it at the cost of a timely report.

With these in mind, we will conduct a whirlwind review of stroke basics before launching into valuable anatomy, stroke syndromes, and some example cases.

1.1 Basics

In the broadest terms, stroke comes in two main varieties: ischemic and hemorrhagic [1].

- Ischemic stroke accounts for approximately 85%.
  - Middle cerebral artery (MCA) distribution infarcts are the most common.
- Hemorrhagic stroke accounts for 15%.
  - Hypertensive rupture of Charcot-Bouchard microaneurysms in the basal ganglia is the most common cause.

1.2 Ischemic Stroke

Ischemic stroke can be further broken down into vascular distributions: anterior circulation and posterior circulation.

- Anterior circulation strokes account for 80% of arterial strokes.
  - Internal Carotid Artery (ICA)
  - Anterior Choroidal Artery
  - Posterior Communicating Artery (PCOM)
  - MCA
  - Anterior Cerebral Artery (ACA)
  - Lenticulostriate Arteries
- Posterior circulation strokes account for 20% of arterial strokes
  - Posterior Inferior Cerebellar Artery (PICA)
  - Anterior Inferior Cerebellar Artery (AICA)
  - Superior Cerebellar Artery (SCA)
  - Posterior Cerebral Artery (PCA)
  - Basilar Artery
- Venous strokes account for up to 3% of all strokes and have some fascinating differences, which are discussed later [2].

1.3 Stroke Basics You Must Understand Before Proceeding

If you are just beginning your foray into neuroradiology, or radiology residency for that matter, I strongly suggest you read the following couple pages and take to heart and mind the material. It will help setup a foundation for your understanding of both the imaging itself and the important literature that guides our path in this field. Yes, I realize it may be a pain in the ass, but you shouldn't run a marathon without doing a few 5Ks first.

For the radiology resident or fellow on call, the following is a common imaging flowchart:

Noncontrast head CT ➔CT angiography of the head/neck & CT perfusion ➔MRI (within 24 hours)

We'll discuss the details later on, but for now you should understand these studies often flow in series. The orders can vary depending on resources, patient claustrophobia, and clinical assessment, but they each add value and are instrumental during patient triage.

If you want to read ahead and get a better understanding as to why, I recommend the 2018 DAWN and DEFUSE 3 trials as enthralling late night reading fare.

We will discuss some items that **you must know** in order to understand stroke literature and communicate effectively with your clinicians. The first is the **National Institutes of Health Stroke Scale (NIHSS)**. It is an 11-item clinical score that has prognostic implications with patients possessing a score of 16 or greater at higher risk of death and those at 6 or less having a better chance of recovery. It is also important to have an understanding of this scale because many studies use these values for inclusion/exclusion criteria and outcomes.

Next is the **Modified Rankin Scale (mRS).** It is a scale of disability after stroke used to predict functional outcomes, and many studies make use of it. Scores 0–2 represent functional independence. Scores 0–3 represent the ability to ambulate independently. Scores of 4 and 5 are indicative of significant disability. And 6, as you guessed, is death. Many of the early landmark stroke studies, and even some recent ones, utilize the **mRS** at 90 days as a primary endpoint measurement. At least one study, SELECT 2, combined scores 5 and 6.

Commit the following to memory, allowing for the fact that definitions and intervention can be controversial.

Large vessel occlusion (LVO) is defined as the occlusion of:

1. Vertebral artery
2. Basilar artery
3. Internal carotid artery
4. M1 segment of the MCA
5. Proximal M2 segment of MCA

Some may argue to include the A1 and P1 segments of the anterior and posterior cerebral arteries, but inclusion of these vessels is controversial. You will never be wrong knowing 1–5 above.

So, if there exist LVOs, then do not also medium vessel occlusions (MeVOs)? Well, as you can infer from the fact that MeVO has now been defined, they certainly do. Unfortunately, their place in stroke management is uncertain as of mid-2025.

We know from good established medical evidence that mechanical thrombectomy is beneficial in LVOs, but two recent trials—DISTAL and ESCAPE MEVO—failed to show the benefit of mechanical thrombectomy in patients with MeVOs.

However, the story is just beginning. Remember, that initial studies in the early 2010s comparing thrombectomy to best medical therapy in patients with LVOs showed no benefit in the thrombectomy arms. **The first published article to do so, MRCLEAN**, was published in 2015, which was followed by four more supporting the use of thrombectomy. The coup de grâce later came with the **HERMES meta-analysis** of these studies, essentially cementing the standard of care.

With respect to MeVOs, as technology and patient selection evolve, study design, outcomes, and ultimately patient management may change.

1.4 CT Aspects

Noncontrast head CT is often the first examination performed on patients suspected of having an acute stroke. It's convenient, very fast, and provides a lot of valuable information. Indeed, this examination is frequently the "gatekeeper" in stroke management, especially if the patient is in the window for chemical thrombolysis.

As it goes, CT imaging can also provide valuable prognostic information and, similar to the mRS and NIHSS, it can be used to generate inclusion and exclusion criteria for studies. This is the essence of ASPECTS, which means Alberta Stroke Program Early CT Score.

It is a brilliantly simple fractionation of the brain into 10 regions. For each region displaying hypoattenuation (cytotoxic edema) in the setting of hyperacute MCA stroke, 1 point is subtracted. Therefore, the lower the score, the larger the stroke. Today, artificial intelligence programs, such as RAPID, can immediately provide an ASPECTS. There also exists a posterior circulation ASPECTS score (Fig. 1.1).

- ASPECTS 0–2: Huge Core. Used as an imaging exclusion criterion in multiple trials, but also as an inclusion criterion in more recent trials.
- ASPECTS 3–5: Large Core. SELECT 2 used this as an inclusion criterion.
- ASPECTS 6–10: Medium/Small Core. DEFUSE 3 used ASPECTS <6 as an exclusion criterion.

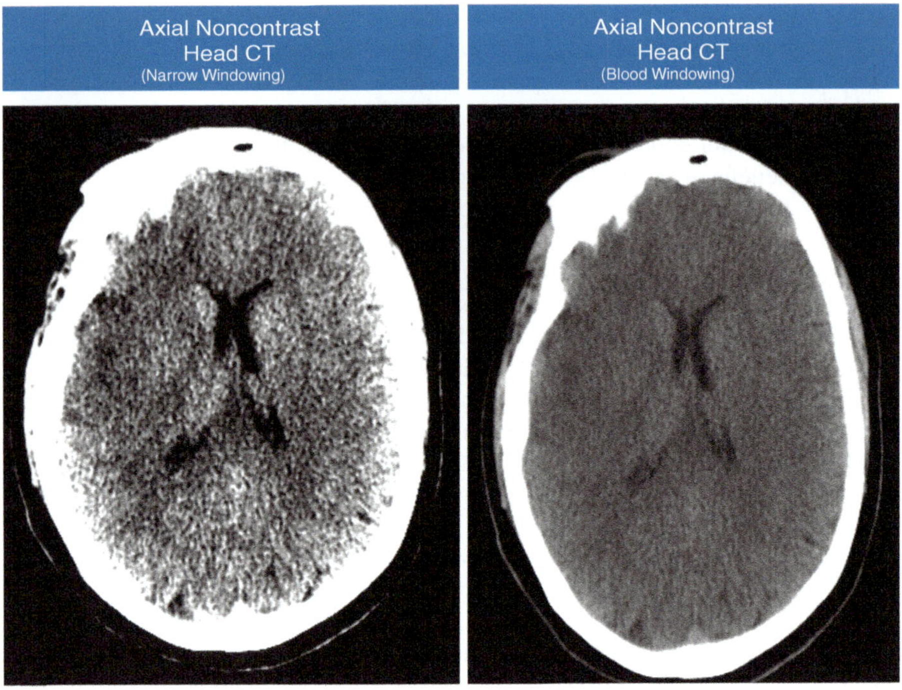

Fig. 1.1 Two axial noncontrast head CT images through the same plane of the brain but in different windows. Notice how the image on your left better shows the right MCA infarct territory. Window and level are essential in discriminating subtle tissue attenuation differences. It is a skill you will build over time, but you should make a habit of reviewing all CT examinations with differing window and leveling. Many PACS systems have the "F" keys as automated window/level hotkeys. Try them out

1.5 Anterior Circulation Strokes (Figs. 1.2 and 1.3, Table 1.1)

There are also some rare congenital variants you should be aware of that connect the anterior and posterior circulations:

1. Persistent trigeminal artery
2. Persistent otic artery
3. Persistent hypoglossal artery
4. Proatlantal artery type 1
5. Proatlantal artery type 2

The persistent trigeminal artery is the most common of these rare variants. It provides an anomalous carotid-basilar anastomosis and may predispose the patient to aneurysm formation. Also, one would exercise extra caution in a patient with this anomaly prior to the Wada procedure. Why?

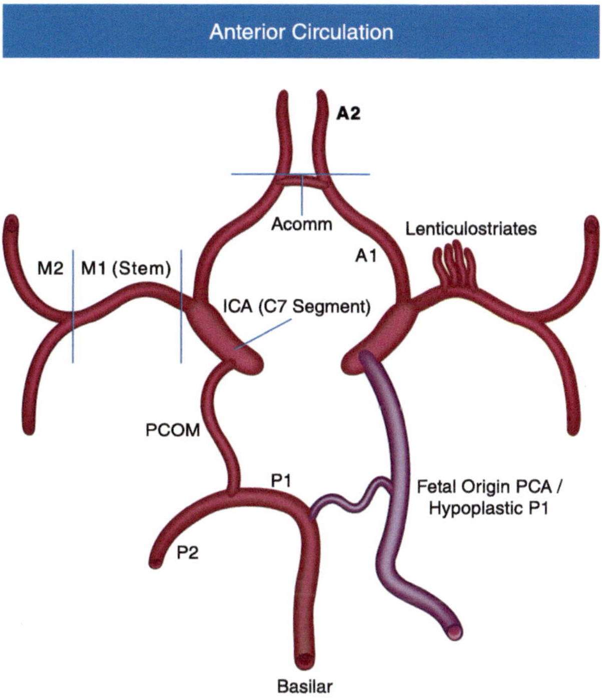

Fig. 1.2 The anterior circulation is depicted here with its connection to the terminal segments of the posterior circulation. The anterior circulation is formed by the ICAs, which branch into the ACAs and MCAs. They are each further divided into descending numeric segments (M1, M2, A1, etc.) The terminal communicating segment of the ICA is known as the communicating or C7 branch. The common variant fetal origin PCA is presented on your right in purple

1.5.1 Bottom Line Up Front

Anterior circulation strokes account for 80% of ischemic strokes [3]. They are the most common because the ICAs supply most of the blood to the brain. Atherosclerotic plaque rupture, cardiac embolism, or embolism from any other source (fat embolism, air embolism, etc.) can obstruct blood flow and cause ischemia. Remember, the brain's metabolic demands are intense. Ischemia and cell death occur rapidly with the most vulnerable cells—Purkinje cells in the cerebellum, and pyramidal cells (especially within select regions of the hippocampus)—dying first (Figs. 1.4 and 1.5).

Five Diagnostic Pearls for the Resident/Fellow on Call:

- Stroke is a clinical diagnosis that often affects the elderly, and there are many stroke mimics. However, you should not allow age bias to influence your reads. Strokes can occur at any age, and it benefits you to read every study with a high index of suspicion.
- Noncontrast CT of the head will be the first study you encounter. The "NCCT HEAD" is the backbone of stroke imaging, just as the lowly chest radiograph is to the thoracic imagers. But remember: there is a reason Felson was able to build a chest X-ray empire on such a simple exam. There is a lot of information packed into the elegant simplicity of both the chest radiograph and the head CT (Fig. 1.6).

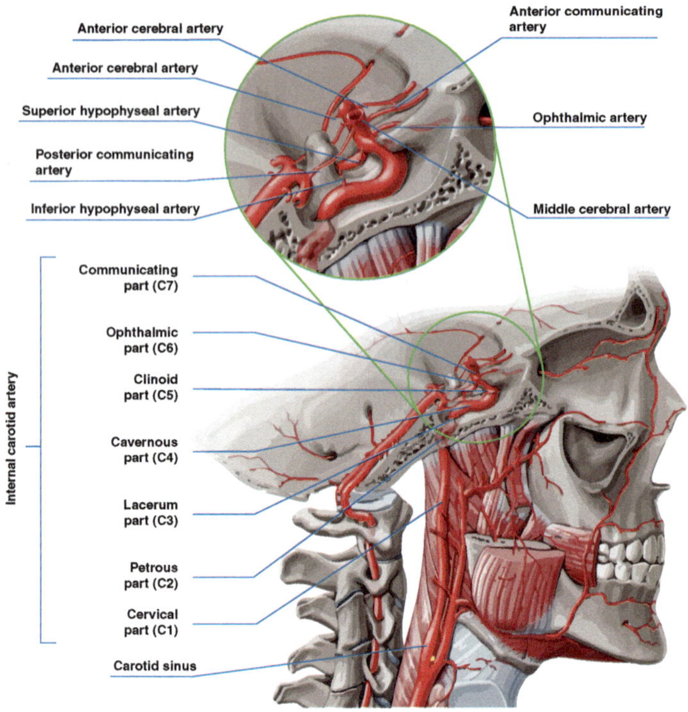

Fig. 1.3 A sagittal illustration of the arteries supplying the head and face. The anterior circulation is labeled in this instance with a magnified inset of the terminal segments of the ICAs. **Notice the C1 through C7 nomenclature. It was established by Bouthillier et al. in 1996 and is commonly used**. (Image courtesy of KenHub)

Table 1.1 In this house, we use the Bouthillier classification system for the ICA! The small branches emerging from C2 and C4 are often not seen on routine diagnostic imaging

ICA segment	Branch arteries
C1: Cervical	None
C2: Petrous	Caroticotympanic, Vidian
C3: Lacerum	None
C4: Cavernous	Meningohypophyseal trunk, inferolateral trunk
C5: Clinoid	None
C6: Ophthalmic	Ophthalmic, superior hypophyseal
C7: Communicating	PCOM, anterior choroidal, MCA, ACA

This may change with the introduction of photon-counting CT and the adoption of 7 Tesla MRI magnets. Be aware that the branches from C4 and their respective subbranches can communicate with segments of the ECA. Memorize the above, and you'll be off to a great start

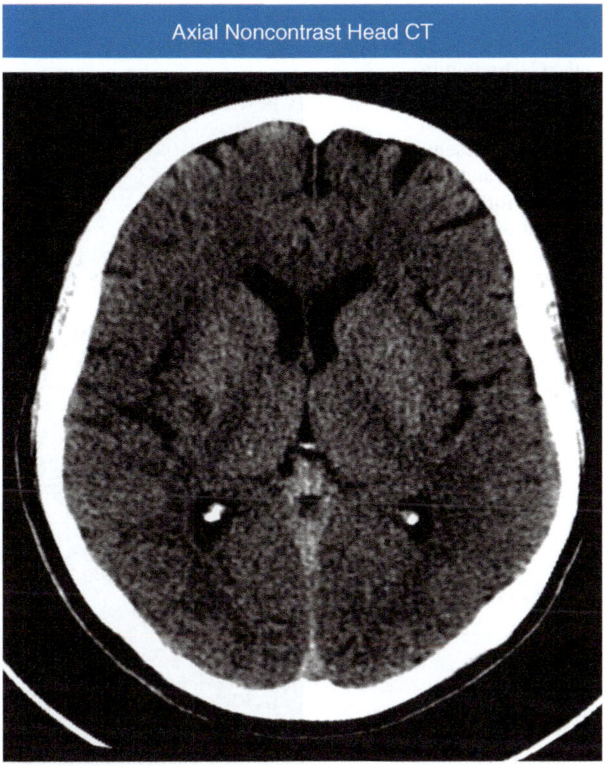

Fig. 1.4 This is an axial noncontrast head CT image through the insula and basal ganglia. Notice the hyperdense deep gray matter, specifically the caudate heads and lentiform nuclei. If you need an anatomy refresher, skip ahead to Figs. 1.16 and 1.17. Here, there is a rounded hypodensity in the right lentiform nucleus. Additionally, there is a loss of gray-white differentiation in the right insular ribbon, indicative of ischemic infarct. Symmetry is your friend. Learn the insular ribbon sign

- This point dovetails the last one. Indeed, it is your responsibility to rule out mass and bleed. However, you can be extraordinarily helpful if you maintain an eye to detail. Is that vasogenic edema as opposed to cytotoxic? Supply a differential. Holy jeeze, there's a mass centered at the petroclival synchondrosis! Make sure it's in your impression. You get the picture? You are responsible for everything in the study.
- The devil is in the details. Describe, describe, describe—you may be helping your senior resident/fellow, or yourself, when the CT angiogram follows. Knowing that the loss of gray-white differentiation is involving not only the insular ribbon and corona radiata *but also* the basal ganglia and superior temporal gyrus will be helpful when hunting for a LVO.
- Use your multiplanar reformations. When you view an examination in three planes, you significantly broaden your horizons. For example, a beam hardening artifact that may blur gray-white differentiation can be confirmed as an unnatural linear band on coronal imaging.

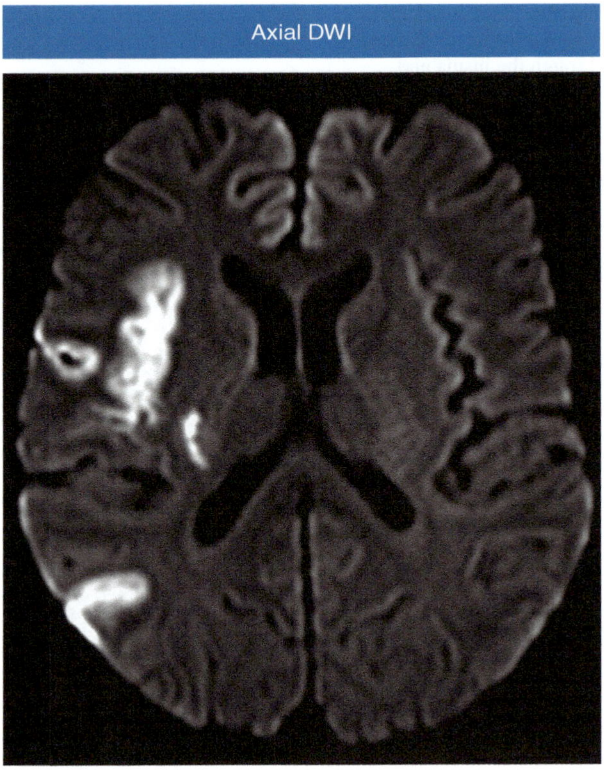

Fig. 1.5 This is the axial DWI image corresponding to the CT finding in Fig. 1.4. Bright signal indicating restricted diffusion is present in the right insula and posterior right lentiform nucleus, though this should be confirmed on the ADC map. The concordance is striking but incomplete. The CT image did not show the right occipital gyriform infarct that you see on the MRI

1. Middle Cerebral Artery

   The MCA distribution stroke is the most common of all strokes. Phylogenetically, the MCA is the youngest of the arteries, which evolved to support the cortical mantle of the massive human brain::flex:: [4]. The clinical manifestations of MCA strokes are variable depending on whether the blockage occurs at the stem (M1) or more distal segments of the artery. It also depends on whether the patient is left or right-hemispheric dominant. Most human beings are left-hemispheric dominant for language and speech, even people who write with their left hand [5]. Furthermore, there is *substantial* variability in MCA vascular coverage of the brain. For the purposes of this discussion, we'll assume "textbook" coverages (Fig. 1.7).

   M1: Also known as the stem, horizontal segment, or the sphenoid segment

   M2: Also known as the insular segment

   M3: Also known as the opercular segment

   M4: Also known as the cortical segment

   Let's occlude the proximal M1 stem segment of the dominant hemisphere, which could affect the following branches and respective territorial supplies.

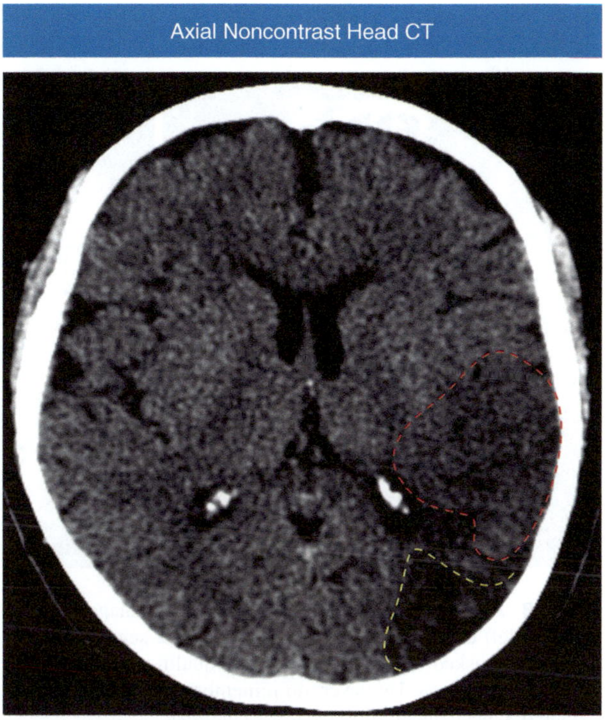

Fig. 1.6 This axial noncontrast head CT has two left-sided infarct regions highlighted with dashes. One is seen more posteriorly in yellow. The other is anteriorly in red. The stroke territory in yellow is more remote with ensuing encephalomalacia. The stroke territory in red is early subacute. Red demonstrates part of the left MCA territory. Yellow is a watershed area

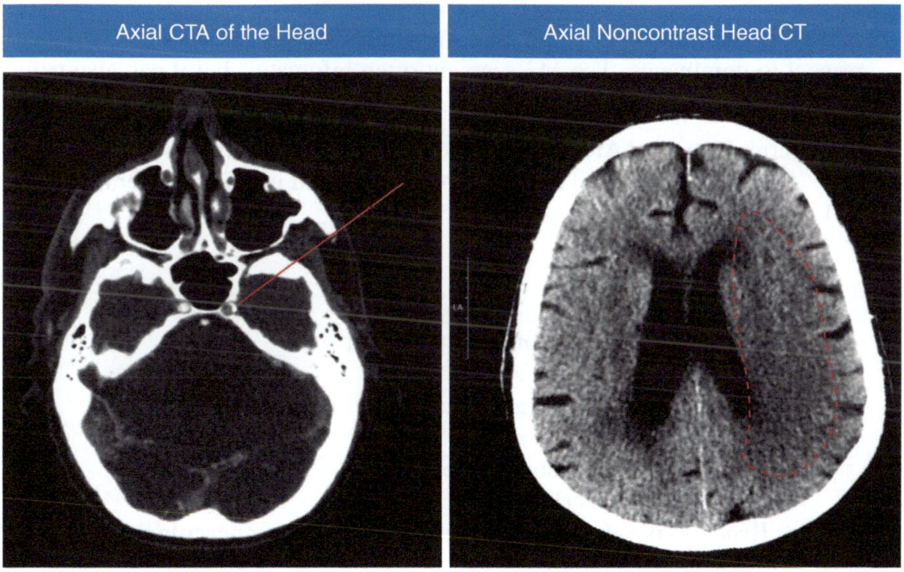

Fig. 1.7 The image on your left demonstrates diminished contrast opacification in the patient's left cavernous C4 segment ICA (red line). The right image shows loss of gray-white differentiation in the patient's left frontoparietal cortex. The area is circumscribed in red hashed lines. Remember that the ICAs contribute to both the ACA and MCA circulations normally. They may also supply the posterior circulation if the patient has a fetal PCA variant

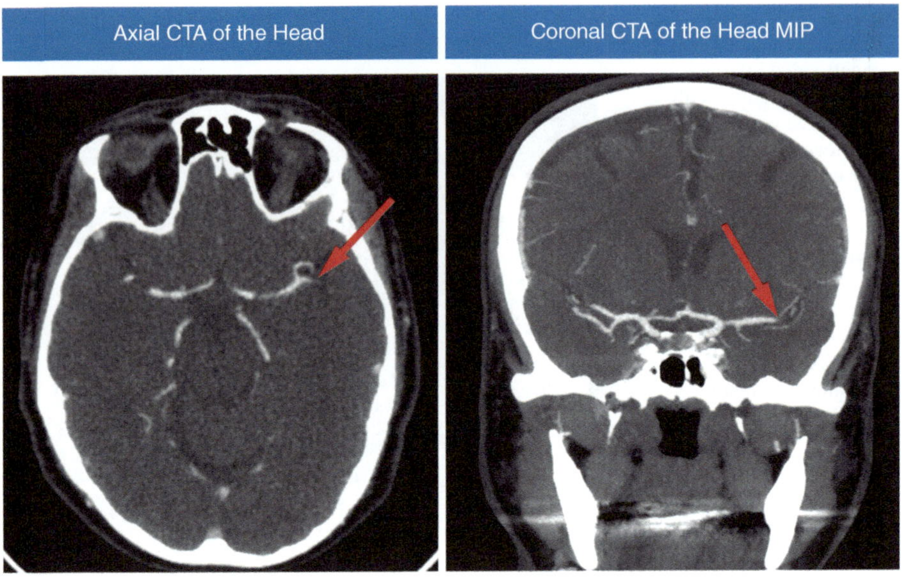

Fig. 1.8 Axial and coronal CT angiographic images demonstrate vascular "cutoffs" of the patient's left distal M1 or perhaps proximal M2 segment MCA (arrows). During your call shifts, you may be asked whether the occlusion qualifies as "large vessel" or LVO in common parlance. In this case, it does. However, do remember that other factors are considered before a patient is declared a mechanical thrombectomy candidate

Again, I stress there is significant variability in the MCA, especially with respect to dominance. The M2 segment often bifurcates, but it can also trifurcate (Fig. 1.8):

- Anterior Temporal Artery (often the first branch from the M1 MCA)
  - Anterior segments of the superior, middle, and inferior temporal gyri
- Lenticulostriate Arteries
  - Caudate nucleus
  - Putamen
  - Globus pallidus
  - Anterior limb and genu of the internal capsule
- Superior Division (M2) MCA
  - Lateral and inferior frontal lobe and anterolateral parietal lobe
    - Primary motor cortex
    - Primary somatosensory Cortex
    - Frontal eye fields (Brodmann 8)
    - **Broca's speech area (pars triangularis/pars opercularis)**
- Inferior Division (M2) MCA
  - Lateral temporal lobe, inferior parietal lobe, lateral occipital lobe
    - **Wernicke's speech area (roughly inferior parietal lobule/Brodmann 22) of the dominant hemisphere**

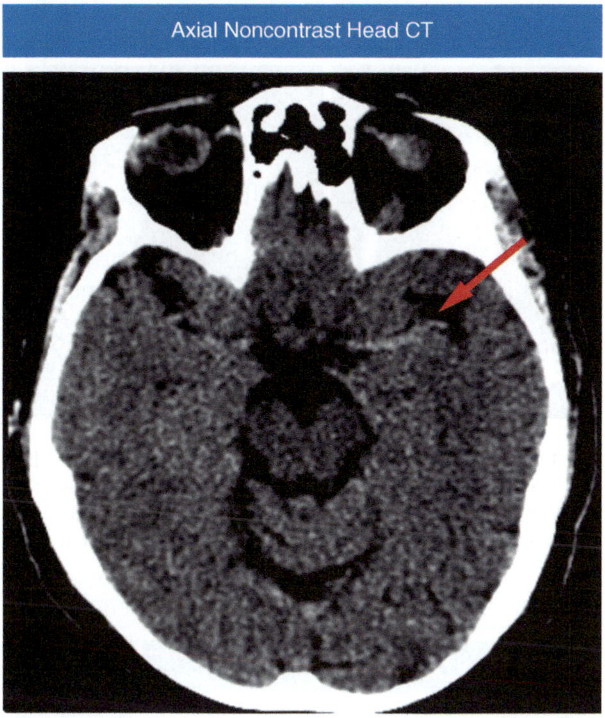

Fig. 1.9 An axial noncontrast head CT slice through the basilar cisterns reveals increased attenuation of the left M1 segment MCA. This can be an early sign of occlusion, but it is also a notorious fake out. Hemoconcentration, patient obliquity in the scanner, and technical parameters can result in false positives. Don't gloss over the finding, but choose your words carefully. In this case, it was a true positive M1 thrombosis

- Optic radiations
- Inferior parietal lobule (angular artery, specifically)

That's a lot of territory. Consequently, one would expect a varied and significant array of clinical symptoms from a proximal M1 stroke. Let's break them down by section (Figs. 1.9, 1.10, and 1.11).

Fun Fact: There is a difference between craniotomy and craniectomy. Use the terms judiciously in your dictations or else you risk losing the confidence of your referring surgeons. Craniotomy is the temporary removal and replacement of the calvarium with fixation. Craniectomy is the removal, which may or may not be followed by mesh cranioplasty.

- Anterior Temporal Artery
  - Anterior segments of the superior, middle, and inferior temporal gyri:
    - Interestingly, this artery doesn't get too much attention in the stroke literature, but it has been studied for arterial grafting [6].
- Lenticulostriate Arteries
  - Caudate infarcts can result in [7, 8]:
    - Abulia (if putamen also involved)
    - Psychic akinesia (athymhormia)
    - Language difficulties if dominant side affected

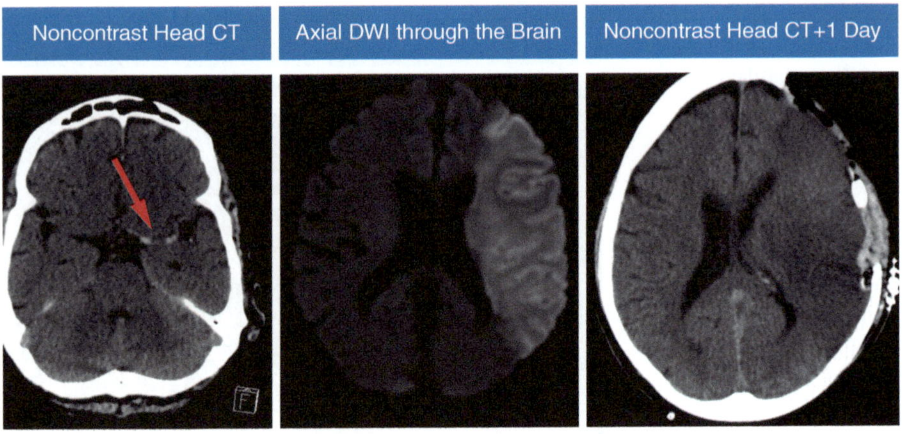

Fig. 1.10 Perhaps a more convincing sequence of images than what Fig. 1.9 showed. On your left, the arrow points toward a hyperdense left M1 MCA. The middle DWI shows a large left MCA-distribution area of restricted diffusion. The right image shows a large territory of cytotoxic edema and mass effect for which the patient underwent left decompressive craniectomy

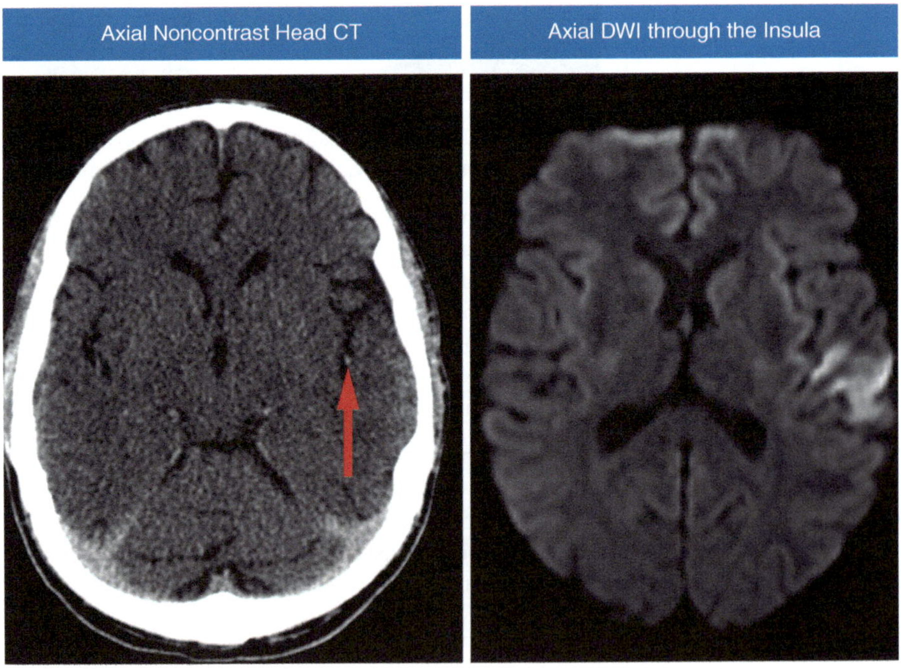

Fig. 1.11 An eye for fine detail can definitely pay dividends in this field, as you see on the annotated noncontrast axial CT through the insula (left, arrow). **This is called the MCA dot sign.** Symmetry is your friend. The subsequent MRI reveals restricted diffusion consistent with an acute infarct in the left M2 MCA distribution

        - Patient difficulty protecting their airway as reflected by increased penetration-aspiration scale (PAS) scores
    - Putamen/globus pallidus, also known as the lentiform nucleus. Isolated lenticular infarcts are very rare [9].
        - Hemiparesis (without movement disorder)
        - Hemisensory loss
    - Posterior limb/genu internal capsule [10]
        - Posterior limb: Contralateral hemiparesis/hemisensory loss/vision and hearing deficits if the retrolenticular fibers are infarcted.
        - Genu: clumsy hand dysarthria, dysphagia (Fig. 1.12).
- Superior Division of the M2 MCA [11]
    - Primary motor cortex
        - Contralateral hemiparesis
            - Upper extremity > Lower (remember your homunculus!)
    - Primary somatosensory cortex
        - Contralateral sensory deficits
            - Face and upper extremities > lower
    - Frontal eye fields: Brodmann area 8 (Fig. 1.13)
        - **Gaze will shift toward the side of the infarct**
            - The injured gaze toward their wounds.
    - Broca's speech area
        - Expressive aphasia (broken speech)
- Inferior Division of the M2 MCA
    - Wernicke's speech Area: Brodmann area 22
        - Receptive aphasia
            - Word salad/nonsensical but fluent speech
    - Optic radiations
        - Homonymous hemianopsia
    - Inferior parietal lobule: Angular gyrus [12]
        - Gerstmann syndrome (Tetrad)
            - Acalculia
            - Agraphia
            - Left-right disorientation
            - Finger agnosia
    - Inferior parietal lobule: Supramarginal gyrus
        - Agraphia

In summation, an unfortunate soul with a proximal M1 occlusion of the dominant hemisphere can present with global aphasia, contralateral hemiparesis and sensory loss, eyes drifting toward the side of infarction, homonymous hemianopsia, confusion, a Gerstmann Tetrad, difficulty protecting their airway, and abulia. And that's just a solid assumption. Every human brain is different, and so is every stroke (Fig. 1.14).

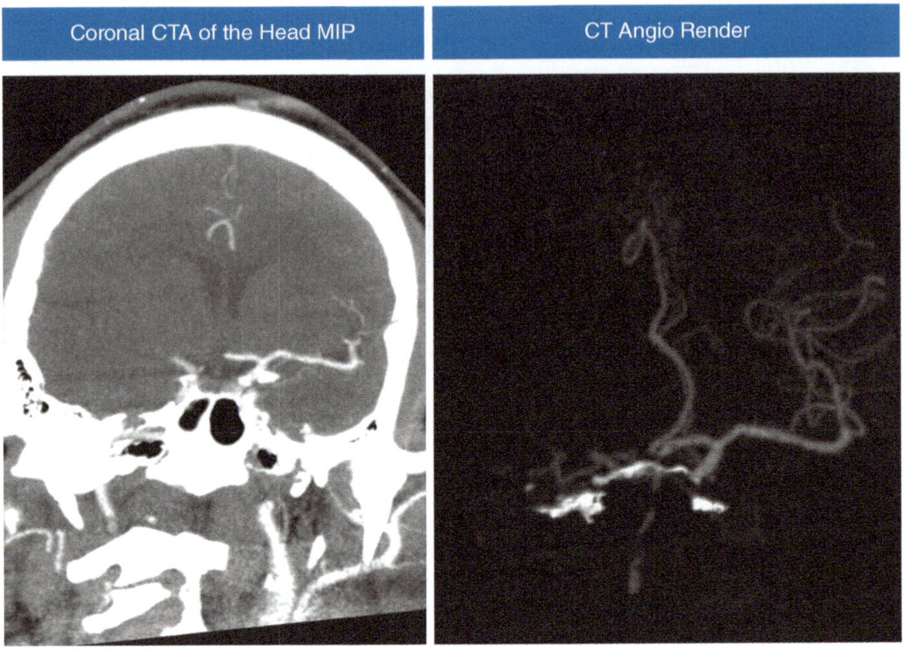

Fig. 1.12 A coronal MIP CT angiographic image is seen on your left and its coronal 3D rendering on the right. Occlusion of the right communicating C7 segment of the ICA results in significant right-sided MCA oligemia. The ACA circulations remain perfused thanks to the anterior communicating artery, even though the right A1 segment ACA is gone

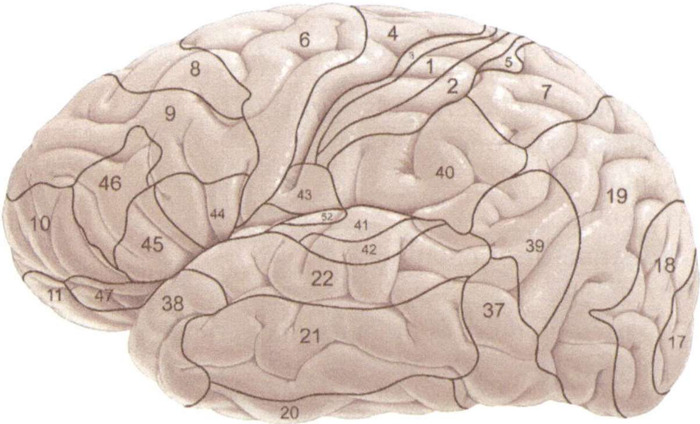

Fig. 1.13 A sagittal illustration of the brain demonstrating topographic Brodmann Areas. For the purposes of stroke identification, familiarize yourself with Broca's Area (BA44/BA45), Wernicke's area (posterior aspect of BA22), the primary motor cortex (BA4), the primary somatosensory cortex (BA1, BA2, BA3), and the inferior parietal lobule (BA39/BA40). The frontal eye fields are BA8. (This figure is courtesy of KenHub)

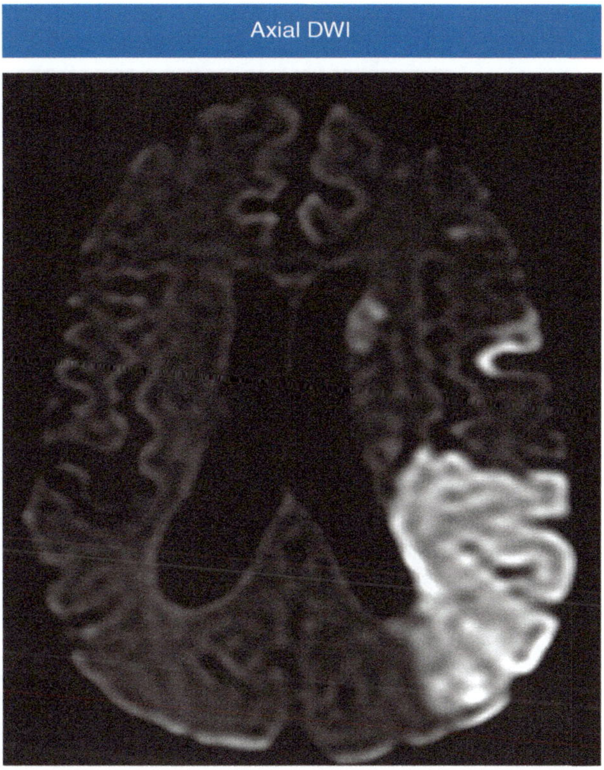

Fig. 1.14 An axial DWI image reveals restricted diffusion in the left parietal cortex, part of the left insula, and left caudate. This is a fitting distribution for a proximal left M1 occlusion

So, what happens when the *nondominant* hemisphere infarcts due to proximal M1 occlusion? A similar result as above with the exception of language. The individual stricken on their nondominant hemisphere will often retain their language abilities. However, their spatial perception will be vastly affected [13, 14]. The "textbook" answer is *hemineglect*, but I believe this fails to completely capture the clinical picture. Here are a few examples of what that means:

- Allochiria: Contralateral neglect of events and sensory stimuli.
- Anosognosia: Lack of awareness of having a disorder or disability.
- Anosodiaphoria: The patient is not concerned about the glaring deficit.
- Unilateral self-neglect: They will shave, clean, and attend to half of their body.
- Somatoparaphrenia: The patient may maintain the delusional belief that the neglected side of their body belongs to someone else (Figs. 1.15, 1.16, 1.17, and 1.18).

2. Anterior Cerebral Artery (ACA)

This stroke is relatively uncommon, comprising around 2% of ischemic strokes [15]. Despite arising from the bifurcation of the C7 segment of the internal carotid artery like the MCA, the ACA cerebral parenchyma can be spared even if there is a complete A1 occlusion. Why? The righteous anterior

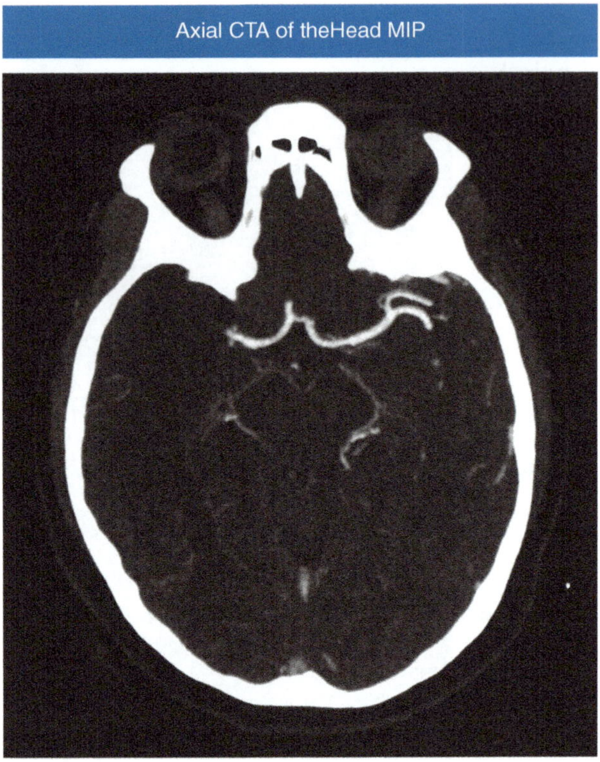

Fig. 1.15 Clearly depicted in this axial MIP through the circle of Willis is a right M1 MCA occlusion as well as developing cytotoxic edema. Observe the relative oligemia of the right cerebral vasculature

communicating artery (Acomm), of course. Regardless, these strokes do still occur, especially distally, due to thromboembolic events. Further, if there is a single stem origin of the ACA—aplastic A1 segment on either side—a bifrontal infarction is in the cards [16]. **Also be aware of the azygos ACA variant, which occurs when the left and right A1 segments fuse into a single A2 trunk.** This finding exists in 1% of the population and is associated with CNS defects and aneurysms.

Again, let's effect a proximal occlusion of the ACA and predict what happens. The nomenclature can be a bit confusing, and it is variable. Medial lenticulostriate arteries can be defined as those arising from the ACA. The lateral lenticulostriate arteries are those arising from the MCA. As always, there is significant variation in topographic vascular coverage. For a very thorough discussion and graphic review, I recommend this article by Kim DE et al. [17] (Fig. 1.19).

- Medial Lenticulostriate Arteries/Recurrent Artery of Heubner
  - Anterior limb of the internal capsule
  - Head of the caudate nucleus (this is debated: some think it's MCA territory)

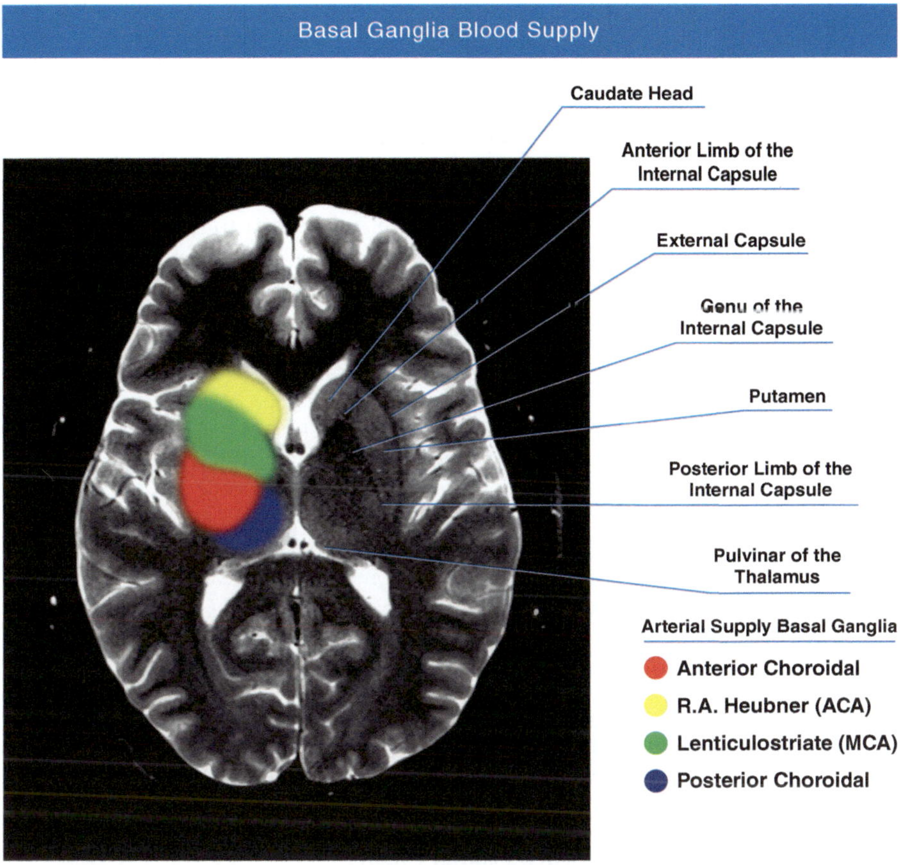

Fig. 1.16 Axial T2 through the basal ganglia demonstrating simplified vascular territorial supply. Note that these territories overlap in vivo and may vary between individuals. The recurrent artery of Heubner (R.A. Heubner) is basically the longest perforating striate artery emerging from the ACA. If I were born 200 years ago, the longest ACA perforator would have been called the *R.A. Hooper*, because I'm positively *basal*. I'll see my way out now

  - Anterior segments of the lentiform nuclei
  - Nucleus accumbens
- The A2 segment length is highly variable, but often the first branch is the orbitofrontal artery
  - Orbital gyri/gyri recti
- Callosomarginal Artery
  - Anteromedial and superior paramedian segments of the frontal lobe
  - Medial precentral gyrus (primary motor cortex)
  - Medial postcentral gyrus (primary somatosensory cortex)
- Pericallosal Artery
  - Corpus callosum through the splenium
  - Anterior cingulate cortex

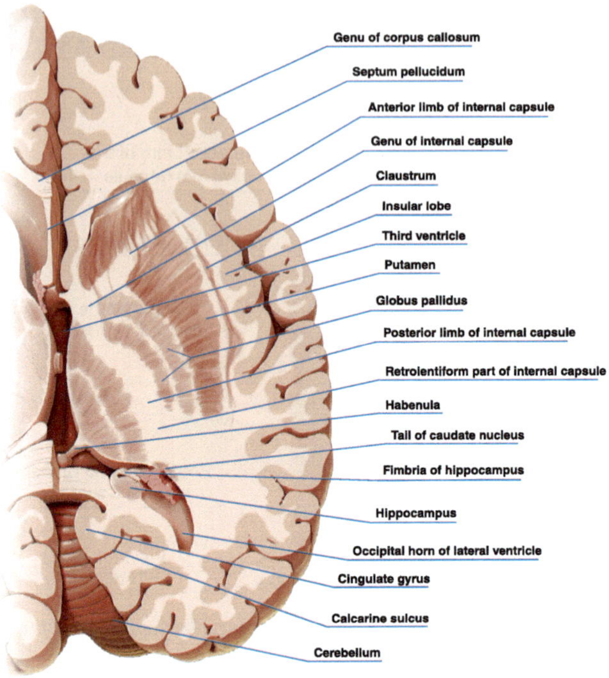

Fig. 1.17 An axial anatomic illustration of the basal ganglia. Learning each and every one of these structures is essential for the neurologist/neuroradiologist. Specific structures and territorial patterns are important not only for stroke localization but also for diagnosing certain toxic/metabolic pathologies and prion illnesses. Image courtesy of KenHub

  - Superior frontal gyrus/paracentral Lobule
  - Supplementary motor area (SMA)

It is here that we will truncate the detail a bit. There are many named branches of the ACA that one can browse elsewhere. Let's focus on the stroke presentations based on what we have.

- Medial Lenticulostriate Arteries/Recurrent Artery of Heubner
  - Anterior Limb of the Internal Capsule [18]
    - Clumsy-hand dysarthria (genu involvement also reported)
    - Behavioral problems
    - Confusion
  - Head of the Caudate Nucleus
    - Abulia
    - Contralateral hemiparesis
    - Emotional problems
    - Declarative and procedural memory problems if left-sided
  - Anterior Lentiform Nuclei Segments
    - Contralateral hemiparesis

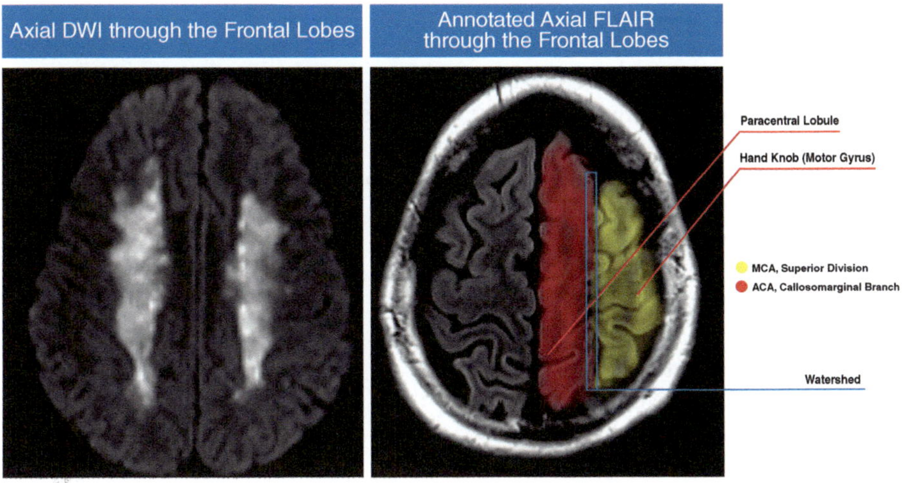

Fig. 1.18 On your left, the MRI image demonstrates restricted diffusion affecting both watershed territories in the setting of a global hypoperfusion event. The image on your right approximates the ACA, watershed, and MCA territories. Note that the "width" of the watershed area, contained within the blue rectangle, varies between human beings

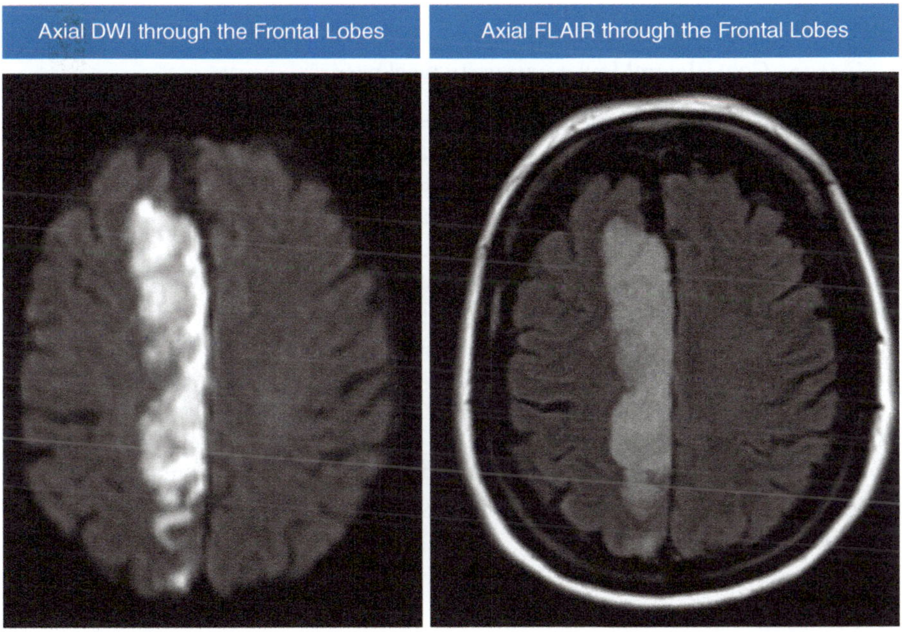

Fig. 1.19 An axial DWI image on your left with a corresponding axial FLAIR image on your right. Note that the abnormal signal in both images is "flush" with the interhemispheric fissure. These findings are consistent with an acute or early subacute right ACA-territory infarct

  - Behavioral problems
- Orbitofrontal Artery (OF)
  - Orbital Gyri/Gyri Recti
    - Functional significance not well described in the literature. Can interfere with ACOM aneurysm clipping. In one report, 6 patients who had an OF artery sacrificed had no neurologic deficits afterwards [19].
- Callosomarginal Artery
  - Primary Motor Cortex
    - Contralateral lower extremity weakness
  - Primary Somatosensory Cortex
    - Contralateral lower body sensory loss
- Pericallosal Artery [20–22]
  - Corpus Callosum
    - Highly variable
      - Subjective cognitive decline
  - Anterior Cingulate Cortex
    - Mutism
    - Apathy
    - Alien hand syndrome
  - Superior Frontal Gyrus/Paracentral Lobule
    - Urinary/fecal incontinence
  - Supplementary Motor Area
    - Contralateral lower extremity weakness

A NCCT head pops up in your list with history "lower extremity weakness." There are myriad causes for this clinical history. You scrutinize carefully the exam, taking note of no hemorrhage and no mass. There's no obvious loss of gray-white differentiation. Shortly thereafter, a CT angiogram of the head and neck arrives along with a phone call from the technologist. The room will be down for a short time for cleaning because the patient abruptly lost control of their bowels and bladder on the table (Fig. 1.20).

Question A: You home in on the anterior circulation while searching for occlusion. If there's no occlusion, what else could it be? What is the next study?

Answer A: Loss of bowel and bladder control can still be the result of infarction. Infarct cannot be ruled out on CTA alone. MRI is required to do this. Loss of bowel and bladder control could also possibly be due to cauda equina syndrome. If this is a diagnostic consideration, MRI of the lumbar spine is appropriate.

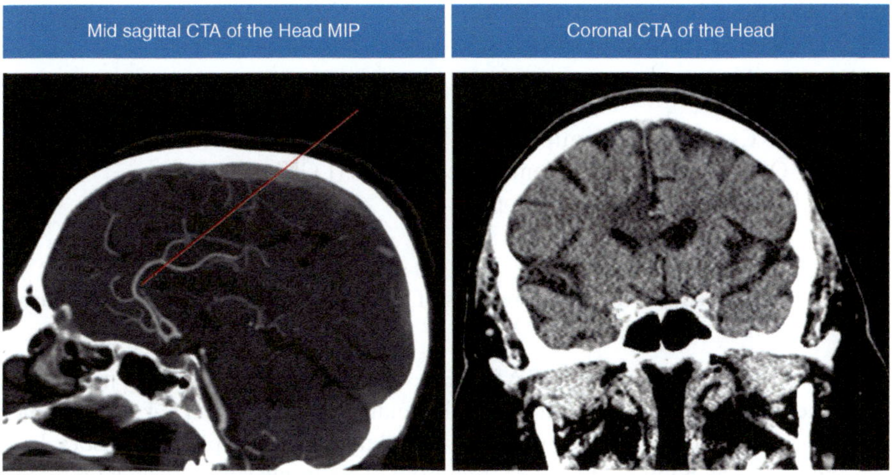

Fig. 1.20 Left annotated (red line) sagittal CT image shows vascular cutoff of the pericallosal A3 segment ACA. Right annotated coronal CT image shows an infarct of the anterior body/genu of the corpus callosum and the cingulate gyrus. What very odd syndrome could this patient later display? Alien hand syndrome, of course. Hopefully, it's not of xenomorph temperament

3. Anterior Choroidal Artery (AchA)

   The AchA arises from the ICA distal to the PCOM. Early in human life, the AchA acts as the "babysitter circulation" of the brain until the vertebrobasilar system—specifically the PCA—develops [23]. There can be some variants where the AchA arises from the M1 MCA or even the PCOM.

   This artery is rarely occluded in isolation, but it covers some important turf.

- Anterior Choroidal Artery Proper
  - Posterior Limb of the Internal Capsule (PLIC)
  - Optic Tracts
  - Lateral Geniculate Nucleus
- AchA Perforators
  - Globus Pallidus Internus
  - Amygdala
  - Medial Temporal Lobe
  - Medial Midbrain

It's rare to get solitary AchA infarcts. Aneurysms are actually the most common pathology of this artery. However, it does cover important territory. One way to remember this is **Wh<u>AchA</u> lookin' at, Bruno?** to reference the immortal and wise George Carlin. The artery supplies the optic tracts and lateral geniculate nucleus of the thalamus as well as the posterior limb of the internal capsule. Infarcts of this structure can result in contralateral hemiplegia, hemisensory loss, and hemianopia [24]. Bruno's fighting prowess would be significantly limited in this scenario.

- Anterior Choroidal Artery Proper
  - Medial Temporal Lobe
    - Memory loss
  - PLIC
    - Contralateral hemiplegia
  - Optic Tracts/Lateral Geniculate Nucleus (LGN)
    - Homonymous hemianopia
    - Quadrantinopia if LGN
  - AchA Perforators
    - There have been case reports of purely cognitive/memory dysfunction, but these are rare.

An MRI arrives in your STAT queue. Opening Bruno's imaging jacket, you notice a lot of old noncontrast maxillofacial, head, and facial CTs. You glance up to the patient's history, which reads, "left-sided hemiplegia, sensory loss, and quadrantanopia." The anterior choroidal syndrome suggests he won't be so pugilistic in the future, at least for a while (Fig. 1.21).

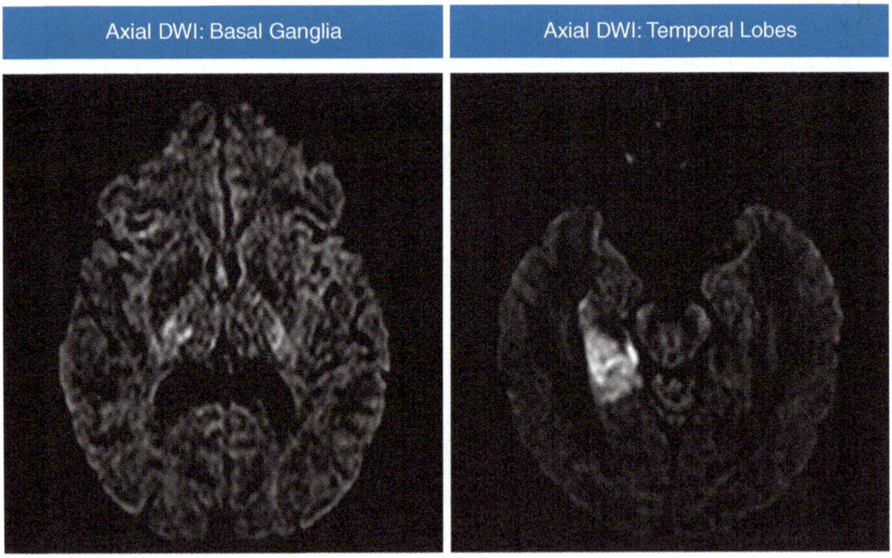

Fig. 1.21 Axial DWI images through the basal ganglia (left) and temporal lobes (right) demonstrate restricted diffusion in the right posterior limb of the internal capsule and medial right temporal lobe

4. Posterior Communicating Artery (PCOM)

   The PCOM is the unsung and highly variable hero of the intracranial circulatory system. This artery allows communication between the anterior and posterior circulations, but it can also serve as a significant supply to posterior brain structures if it levels up and develops into a "fetal origin" PCA (see Fig. 1.2). But wait—there's more! Deep perforating vessels arise from PCOMs, which are better seen on DSA and MRA at 7 Tesla field strength, contributing to deep brain structure collateral supply [25].

   The PCOM is rarely infarcted in isolation, but misadventures can occur if there is a PCOM aneurysm that needs surgical attention. With respect to the PCOM aneurysm, this disease process is a favorite question, as it is focal and extremely important to diagnose. About one in five patients with a PCOM aneurysm will present with oculomotor nerve palsy [26]. Some will have pain behind the eye.

   - Tuberothalamic Arteries
     - Anterior thalamus
   - Pre-mammillary Artery
     - Subthalamic nucleus
     - Hypothalamus
     - Mammillary bodies

   As I said, a de novo infarction of the PCOM is rare. However, territorial infarctions, specifically of the tuberothalamic and AChA arteries, can occur if a PCOM aneurysm is clipped or coiled.

   - Tuberothalamic Arteries [27, 28]
     - Anterior thalamus
       - Temporal disorientation
       - Amnesia
       - Difficulty learning
       - Apathy
   - Pre-mammillary Artery [29, 30]
     - Subthalamic nucleus
       - Contralateral hemiballism
     - Hypothalamus
       - Horner syndrome (first order)
         - Ptosis
         - Miosis
         - Anhidrosis
       - Somnolence
     - Mammillary bodies
       - Amnesia

Questions: Ms. Regina Poppins presents to the ER after waking up at her sister's house with double vision and extreme pain behind her left eye. You—the transitional year radiology resident doing your time on the ER rotation—quickly notice that the patient cannot adduct her left eye past midline. Furthermore, her left pupil is dilated and nonreactive to light. What's the next best study? What territory is at risk during treatment?

Answers: CT or MR angiography of the head. Infarcts involving the thalamus and limbic structures may evolve after PCOM aneurysm clipping. Be aware that cavernous sinus thrombosis can also cause multiple cranial nerve palsies. For bonus points, a PCOM aneurysm will **first affect the parasympathetic fibers of the oculomotor nerve** before the motor portion due to their dorsomedial positioning. As a result, motor function may still be early intact. I know, the pieces fit.

1.6 Posterior Circulation Strokes (Fig. 1.22)

1.6.1 Bottom Line Up Front

These strokes account for 20% of all ischemic strokes and are more prone to misdiagnosis compared to their anterior brethren [31]. For this discussion, we will begin inferiorly at the level of the medulla and proceed superiorly. In doing so, we can utilize our "PASP" mnemonic to solidify our understanding of the anatomy, even though some additional arteries not included in the mnemonic may be involved. PCA strokes will be included in this section, but be aware of the so-called fetal origin PCA variant.

Five Diagnostic Pearls for the Resident/Fellow on Call:

- Be wary of any patient who presents with acute dizziness. Most patients presenting with dizziness have benign positional vertigo, alcohol intoxication, or other myriad benign causes. However, you must not let a common symptom jade your assessment of a lethal pathology. Yes, brainstem strokes are the most lethal type of stroke [32].
- Noncontrast CT of the head is often the first imaging obtained. Unfortunately, sensitivity for detecting posterior fossa strokes is around 40%—not great. This is due to beam hardening and the streak artifact. Keep an eye out for subtle blurring of gray-white differentiation, intra-axial/intraventricular neoplasms, and a dense basilar artery.
- The basilar artery is considered a large vessel for "large vessel occlusion" (LVO) targeted mechanical thrombolysis. CT angiography is best for evaluating the vertebrobasilar confluence and basilar artery proper after a heaping 100 mL IV bolus of iodinated contrast, assuming the patient is not allergic to iodinated contrast. If absolutely contraindicated, then gadolinium-enhanced or time-of-flight (ToF) MR angiography is the next best thing.

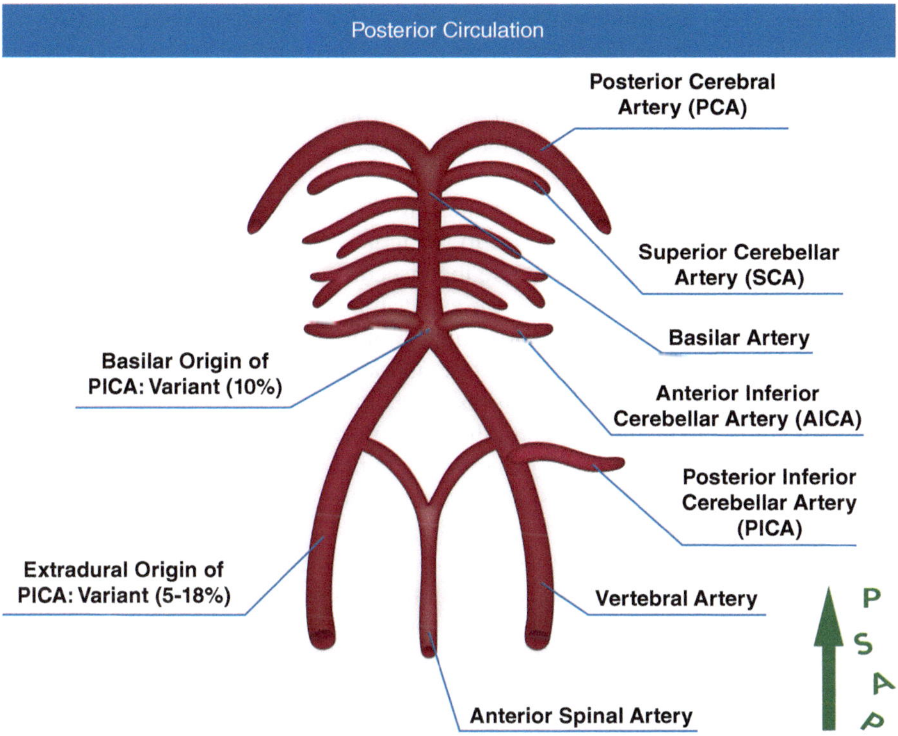

Fig. 1.22 A coronal illustration of the posterior circulation demonstrating a common configuration. It is formed by the confluence of the two vertebral arteries, though it is not uncommon for the vertebral arteries to be variable in size. From inferior to superior, remember the mnemonic PASP for (PICA, AICA, SCA, and PCA)

- If MRI is on the menu, carefully cross-check the DWI and ADC sequences with FLAIR. DWI signal can be spuriously elevated close to bone-soft tissue interfaces (susceptibility artifact). **Be wary of increased DWI signal in the brainstem.** White matter tracts can show mild DWI signal hyperintensity normally.
- Don't let anybody "crack" your neck. High velocity low amplitude manipulation of the cervical spine is a risk for vertebral artery dissection (VAD). And a majority of VADs will go on to ischemic infarct [33]. If a patient is sent to MRI with concern for dissection, you must add an Axial T1 sequence with fat saturation from the aortic arch through the Circle of Willis. **Be wary of increased T1 signal in the V3 segment of the vertebral artery.** The adjacent venous plexus can be bright, thus causing a possible false positive!

Before we begin studying the posterior fossa stroke syndromes, I'd like to introduce a wee bit of embryology that helps explain the *why* behind the brainstem and spinal cord motor and sensory distributions.

Fear not, I'll be brief.

In short, there are three motor modalities in the brainstem: somatic, visceral, and branchial. There are also three sensory modalities in the brainstem: somatic, visceral, and special.

The dorsal structures of the spinal cord are formed by the alar plate, while the ventral structures are formed from the basal plate. They are divided by the sulcus limitans.

The dorsal structures give rise to the sensory components of the spinal cord, while the ventral structures give rise to the motor. Furthermore, the somatic sensory structures are more dorsal, while the visceral sensory structures are closer to the sulcus limitans. When you cross the sulcus limitans, you will then next encounter visceral motor neurons and then somatic motor neurons.

With respect to the brainstem, the orientation of the basal and alar plates after maturation will be different. The alar plate structures will move laterally. The basal plate structures will be medial. Thus, the motor nuclei will be more midline and the sensory lateral. The otic placode will give rise to the vestibular and cochlear nuclei, which will remain lateral to the sulcus limitans (they'll hang out with their sensory buddies).

There is a logical organizational structure in the craniocaudal plane as well.

Medial to the sulcus limitans will be motor. Lateral will be sensory. Within the medial motor column, the somatic motor nuclei will be stacked one on top of the other. The motor components of cranial nerves 3 and 4 will be the midbrain, CN6 will be the pons, and CN12 will be the medulla. Just lateral, you will see the branchial motor nuclei (CN5, CN7, CN9, CN10). These exist in the pons and medulla (NOT the midbrain). Just a little further, you will get the visceral motor (CN3, CN9, and CN10 with the Edinger-Westphal Nucleus functioning as the parasympathetic fibers of CN3).

As you can see, there is method to the madness of cranial nerve organization in the spinal cord and brainstem. If you simply form an archetype in your mind of "lateral sensory/medial motor" in the brainstem and "ventral motor/dorsal sensory," it will go a long way in localizing strokes when viewing MRIs. The somatotopic organization can be a bit more difficult given the limitations of MRI special resolution (Fig. 1.23).

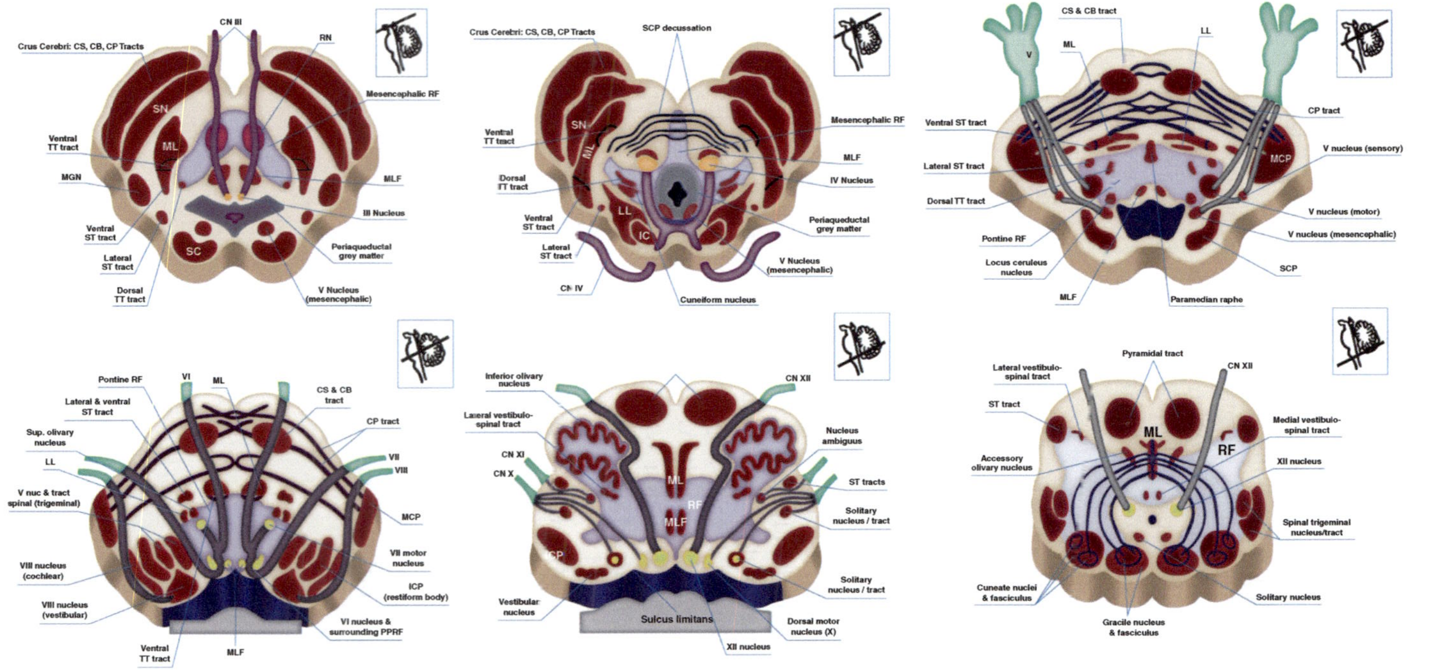

CB = corticobulbar, CP = corticopontine, CS = corticospinal, IC = inferior colliculus, ICP = inferior cerebellar peduncle, LL = lateral lemniscus MGN = medial geniculate nucleus, ML = medial lemniscus, MLF = medial longitudinal fasciculus,MCP = middle cerebellar peduncle, PPRF = paramedian pontine reticular formation, RF = reticular formation, RN = red nucleus, SC = superior colliculus, SCP = superior cerebellar peduncle, SN = substantia nigra, ST = spinothalamic, TT = trigeminothalamic

Fig. 1.23 A superb and rather busy image depicting the topography of the midbrain, pons, and medulla. Get plenty of miles out of this figure. It will also be used in specific stroke instances. (Adapted with permission from Kim et al. [34])

* The midline somatic motor nuclei and tracts can be thought of as frequently and totally under your control. Additionally, just remember that you can divide 12 evenly by 3, 4, 6, and 12 and you have your somatic motor cranial nerves.
* Cranial nerves 9 and 10 possess special visceral efferent (SVE) fibers that supply the pharyngeal arches. Thus, damage to these structures in the nucleus ambiguous will result in difficulty in speaking and swallowing.
* The palatoglossus muscle is the only tongue muscle not innervated by CN12.

I highly recommend ***The Noted Anatomist***, Dr. David A. Morton's YouTube Channel for exquisite explanations of the sulcus limitans, brainstem anatomy, and cranial nerve nuclei. Specifically, the following two videos are well worth your time:

1. Brainstem. Motor nuclei https://youtu.be/Y5wbD8xtUJU?si=_ZEyYEPS1fq5uHEP
   And
2. Foundational features of the brainstem https://youtu.be/BAfrD6pf22k?si=Wts3-7VXsecscle1

1. Posterior Inferior Cerebellar Artery (PICA)
   This artery is important and highly variable. It frequently emerges from the V4 segment of the vertebral artery but can also frequently emerge elsewhere, including the basilar artery. It can be aplastic or hypoplastic. And the vertebral artery itself can terminate as the PICA. PICA is also an ornery artery, implicated in neurovascular compression syndromes, arteriovenous malformations, and 40% of cerebellar infarcts [35].
   - Medullary Branches
     – Dorsal and Lateral Medulla
       - Nuclei of CN 8, 9, and 10, including nucleus ambiguus
       - Spinothalamic tract
       - Spinal trigeminal nucleus
       - Inferior cerebellar peduncle
       - Sympathetic fibers
       - Tonsillar and Cortical Segments

  - Cerebellar Hemisphere
    - Posterior inferior hemispheric parenchyma
    - Inferior vermis

The PICA supplies some important territory with textbook recognizable clinical manifestations if the disease and patient are reading the textbook. Often, they do not. The stakes are quite high here, as the posterior fossa is a tight space that does not accommodate swelling brain. Worse, if this stroke is missed and the patient goes home with Zofran and Tylenol, they might not wake up. Let's occlude it proximally (Figs. 1.24, 1.25, and 1.26).

- Medullary Branches
  - Dorsal and lateral medulla (Lateral Medullary/Wallenberg's Syndrome)—this is the most clinically relevant of the brainstem strokes [36].
    - Vestibular nucleus (CN8)
      - Vertigo
    - Nucleus ambiguus (CN9, 10, and some 11)
      - Dysarthria
      - Dysphagia
      - Hiccups [37]

      (i) Specific but not pathognomonic

        (1) **Other diseases affecting the area postrema (like Neuromyelitis Optica) can also cause hiccups** [38]

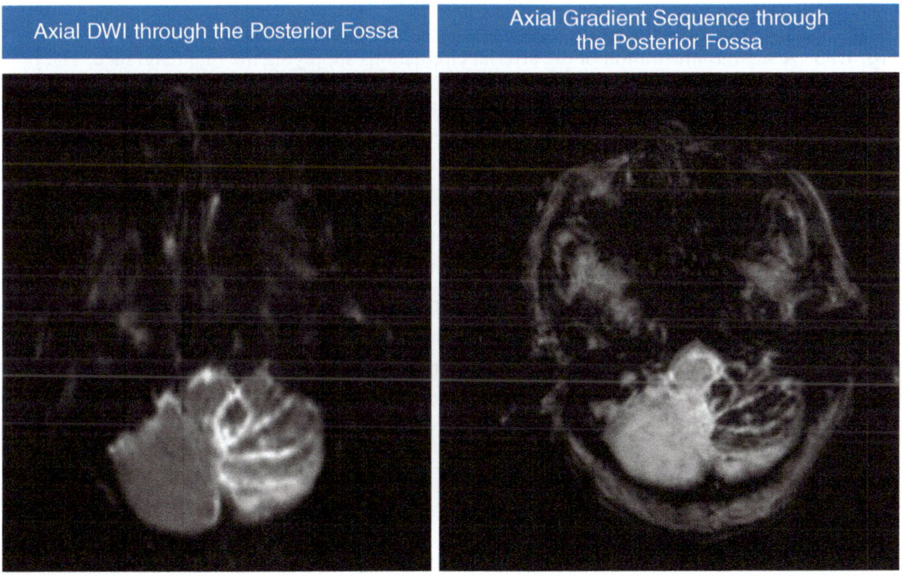

Fig. 1.24 On your left, we have an axial ADC image through the posterior fossa demonstrating decreased left cerebellar diffusivity consistent with an infarct. On the right, an axial gradient image of the same level. Gradient "bloom" along the folia of the left cerebellar hemisphere is consistent with hemorrhagic transformation

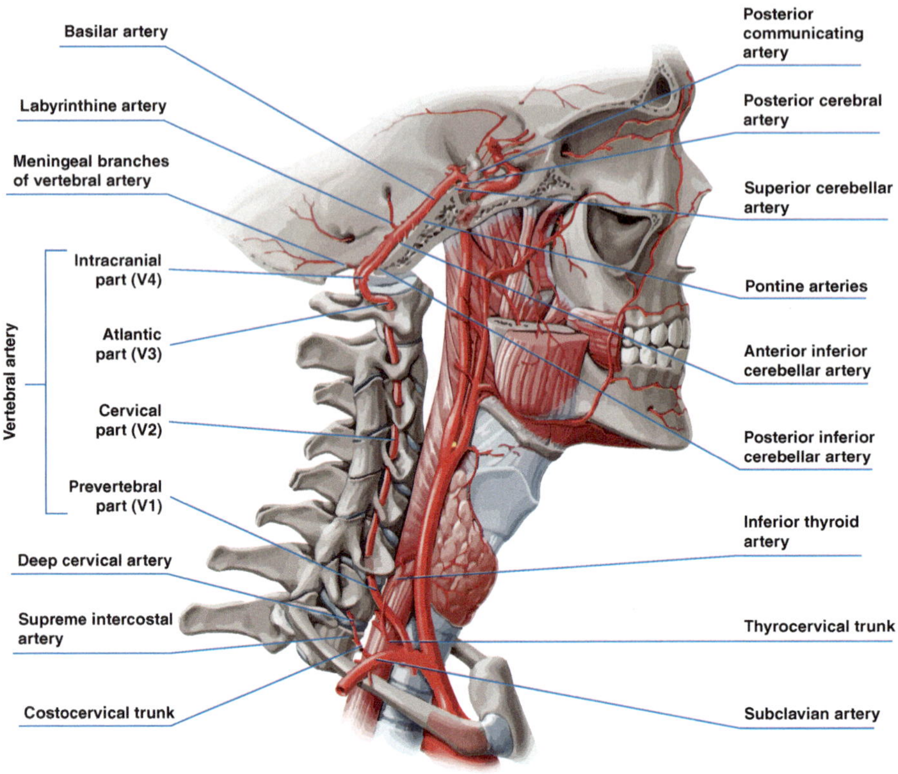

Fig. 1.25 A sagittal illustration of the arterial circulations supplying the brain and face. The posterior circulations are labeled in this instance. The vertebral arteries are segmented as V1 through V4. When describing the vertebral arteries on MR or CT angiograms, it is good practice to include their origins (commonly subclavian) and dominance if one side is larger than the other. **Also, never forget to look carefully at the foramina transversaria.** More on that later. This image is courtesy of KenHub

    - Spinothalamic tract
      - Contralateral loss of pain and temperature sensation in the body.
    - Spinal trigeminal nucleus (CN5)
      - Ipsilateral loss of pain and temperature sensation in the face.
    - Inferior cerebellar peduncle
      - Ipsilateral hemiataxia
    - Sympathetic fibers
- Ipsilateral Horner's syndrome
- Tonsillar and Cortical Segments
  - Cerebellar Hemisphere/Cerebellar Tonsils
    - Nonspecific
      - Headache
      - Nausea

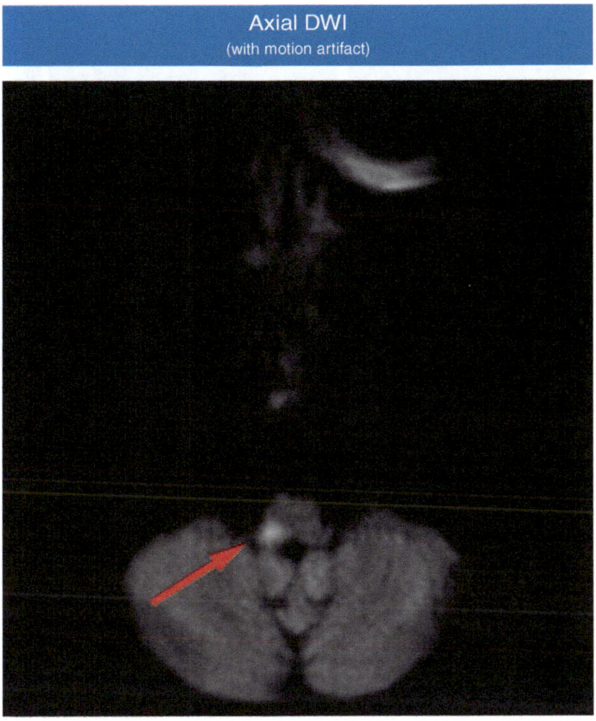

Fig. 1.26 Axial DWI shows restricted diffusion in the right posterolateral medulla (arrow). Be aware that DWI/ADC imaging through the posterior fossa is prone to artifact. Symmetry and comparison studies will be of exceptional use when evaluating this area

  - Vomiting
  - Vertigo
- A little less nonspecific
  - Ataxia
  - Hand incoordination

While reading your tenth consecutive CT angiogram of the head and neck, ☹, you take note of yet another nonspecific clinical history: vomiting. The patient's noncontrast head CT shortly before was completely normal, but the right vertebral artery looks corrugated and beaded through its V3 segment. Anatomic variant? FMD? Contrast bolus? Atheromatous plaque? You call the beleaguered ER attending, who adds that the patient nearly crashed their car while driving home from a chiropractic appointment.::lightbulb:: Do go forth, good doctor, and suggest an MRI with the finest diffusion-weighted and axial fat-saturated T1 fixins' (Fig. 1.27).

If, instead, you have a 2D time-of-flight (TOF) reconstruction of the patient's head and neck vascular anatomy, do be aware that when the images are stitched together, you may see something called **stair step artifact**. This can simulate pathology, so always cross-check findings on your source imaging.

2. Anterior Spinal Artery (ASA)

Formed by the confluence of two small vessels emerging from the medial distal V4 segment of the vertebral arteries, the ASA runs the ventral length of

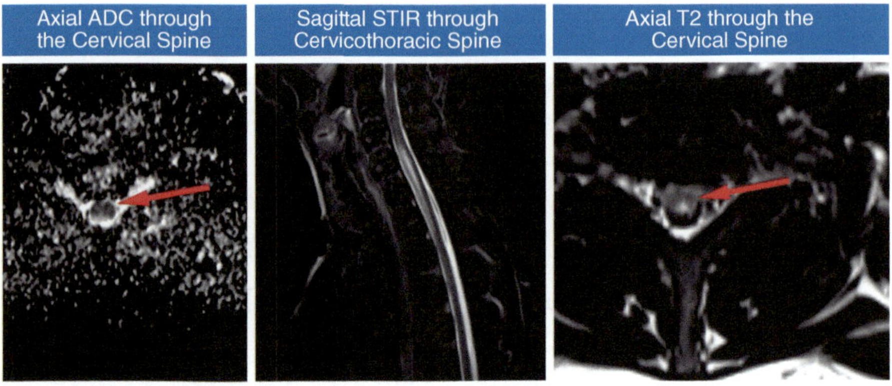

Fig. 1.27 Here is something of exceptional rarity: decent diffusion-weighted imaging of the spinal cord. The far-left image shows signal hypointensity consistent with decreased diffusivity within the anterior spinal cord. This finding corresponds to longitudinally extensive STIR signal abnormality on sagittal imaging and axial T2 signal hyperintensity. These findings support an anterior spinal artery cord infarction. The patient presented with rapidly progressive quadriparesis

the spinal cord but also supplies the hypoglossal nerve, the medullary pyramids, and the medial lemniscus. The artery supplies the anterior two-thirds of the spinal cord with medullary and radicular branches, and occasional anastomoses via fine pia collaterals. Farther caudally, the anterior spinal artery will join with the Artery of Adamkiewicz, which usually arises from the descending aorta between T8 and L2. **The Artery of Adamkiewicz is a favorite pimp question of wily attendings, but more importantly, damage to this artery during surgery can lead to anterior spinal cord infarcts**. Furthermore, hyperextension of the spine, such as during surfing or yoga, can precipitate cord infarcts after damage to the artery of Adamkiewicz.

- Corticospinal Tract
- Corticobulbar Tract
- Medial Lemniscus
- Hypoglossal Nerve Fasciculus
- Spinothalamic Tract
- Vestibular Nuclei
- Medial Longitudinal Fasciculus (MLF)

Atherosclerotic disease is the major causative factor of the medial medullary syndrome, aka Dejerine syndrome. So, we won't be tossing emboli, but rather simply waiting for the patient to ingest another triple cheeseburger (admittedly delicious—no judgment here). When the infarct does occur, it most commonly occurs in the rostral ventral segment of the medulla, accounting for the predilection for motor presentation [39].

- Corticospinal Tract
  - Contralateral weakness

- Corticobulbar Tract
  - Dysphagia/dysarthria
- Medial Lemniscus
  - Contralateral loss of body fine touch and vibratory sensation
- Hypoglossal Nerve Fasciculus
  - Ipsilateral tongue deviation (rare)
  - Clumsy tongue movements
- Spinothalamic Tract
  - Contralateral loss of body pain and temperature sensation
- Vestibular Nuclei
  - Vertigo
- MLF
  - Oculomotor dysfunction

A normal noncontrast MRI of the brain flies from your queue with the press of a button, but *something* doesn't sit quite right with you. After moments of pensive deliberation, you shrug and eye the next case. However, a cold chill sets over your shoulders, and you first decide to refill your coffee. But upon standing, you're confronted by an otherworldly apparition emitting a dull blue aura. It's Dejerine himself—the ghost of infarct past—and he does not look pleased. The ornery dead Frenchman sticks his tongue out at you, which notably deviates to the right, before vanishing. *Perhaps that punctate DWI signal hyperintensity in the right ventral medulla was real*? (Fig. 1.28)

3. Anterior Inferior Cerebellar Artery (AICA)

The AICA emerges from the basilar artery in most instances, but rarely can it take off from the vertebral artery. This little fellow will collateralize with the PICA, so there can be shared circulation. It is involved in 1% of ischemic strokes. The more distal labyrinthine artery arises from the AICA.

- Pontine Branches
  - Lateral pons
    - Trigeminal nerve
    - Abducens nerve
    - Facial nerve
    - Vestibulocochlear nerve
    - Spinothalamic tract
    - Sympathetic tract
    - Paramedian pontine reticular formation (PPRF)
  - Brachium pontis
    - Pontocerebellar tracts
- Flocculopeduncular and Cortical
  - Flocculus and nodulus (nodulus moreso by PICA)
  - Anterior inferior cerebellar hemisphere

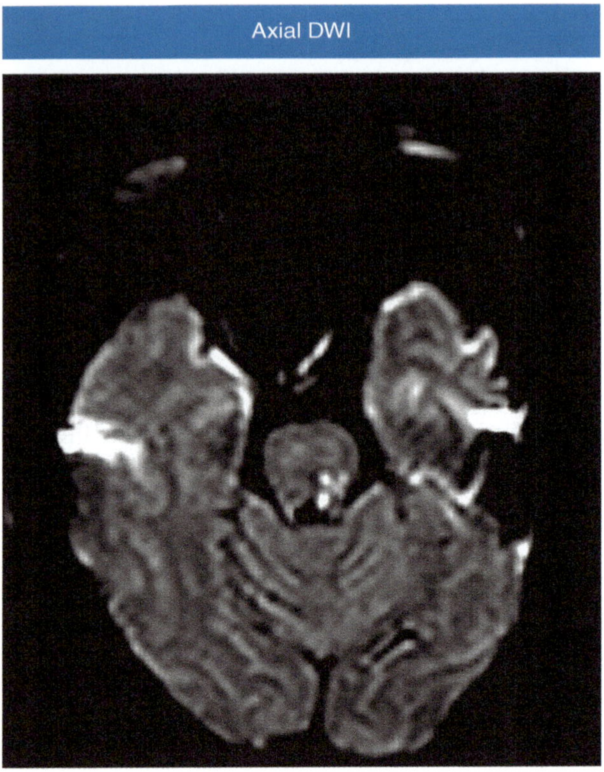

Fig. 1.28 An axial DWI image through the pons shows restricted diffusion in the left posterolateral pons. The patient presented with vertigo, ataxia, nausea, left-sided facial weakness, hearing loss, and loss of pain sensation on the contralateral side of the body. Marie-Foix—a tough act to follow! You will also see the signal hyperintensity along the lateral margins of the temporal lobes. This is spurious and secondary to the bone susceptibility artifact

- Labyrinthine Artery
  - Superior labyrinth
  - Cochlea

Many anastomoses exist between the AICA and PICA. Therefore, it can be unusual to see a textbook territorial infarct. However, there are a few pearls one can clutch in order to bend one's suspicions towards an AICA infarct. For one, vertigo is the most common presenting symptom with AICA infarcts [40].

- Pontine Branches
  - Lateral Pons
    - Trigeminal Nerve (V)
      - Ipsilateral facial loss of pain and temperature sensation
      - Weakened mastication (CNV3)
    - Facial Nerve (VII)
      - Ipsilateral facial muscle paralysis
    - Vestibulocochlear Nerve (VIII)
      - Vertigo
      - Central hearing loss
    - Spinothalamic Tract

      - Contralateral loss of pain/temperature sensation
    - Sympathetic Fibers
- Ipsilateral Horner's syndrome
  - PPRF
    - Horizontal gaze palsy
  - Brachium Pontis [aka Middle Cerebellar Peduncle]—Afferent white matter tracts abound via the corticopontocerebellar (CPC) pathways
    - CPC
      - Ipsilateral ataxia
      - Tremor
      - Scanning speech
  - Labyrinthine Artery
    - Superior Labyrinth/Cochlea
      - Acute hearing loss
      - Nausea
      - Vomiting
      - Vertigo

AICA and PICA are close. They share anastomoses and some territory. As such, their respective stroke syndromes can have significant overlap. To confuse matters further, the middle cerebellar peduncle is considered a watershed zone between AICA and SCA. Marie-Foix is the eponym given to lateral pontine syndrome. One way I remember the difference is:

Wallenberg, Wallenberg —quite a word to swallow (dysphagia/dysarthria/hiccups).

But Marie-Foix Syndrome is a tough act to follow (can't follow if you're ataxic and deaf).

4. Superior Cerebellar Artery (SCA)

There isn't much variation with this artery, except for the occasional duplication. I find the SCA/PCA duo useful to orient when viewing coronal angiograms. Like the PICA, this artery can irritate its neighbors, specifically the trigeminal nerve [41].

- Proximal Perforators
  - Midbrain
  - Pons
  - Tectum
  - Pineal body
- Rostral and Caudal Branches
  - Vermis
  - Superior hemisphere

Superior cerebellar artery strokes are nonspecific. "The literature" breaks them down into proximal and distal. So, we will discuss them in a consolidated manner, which I hope will help you, dear reader, from getting lost in the sauce (Fig. 1.29).

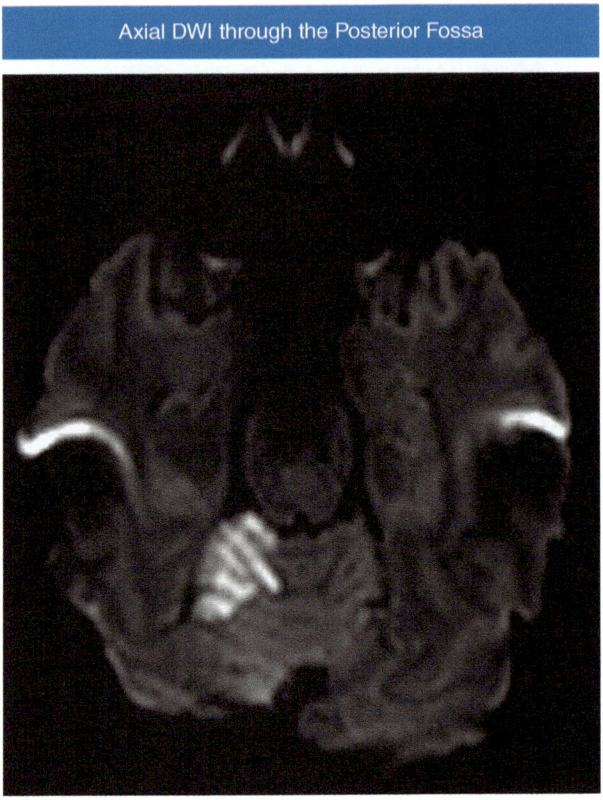

Fig. 1.29 Restricted diffusion is seen in the superior right cerebellar hemisphere. The patient's symptoms were nonspecific and included nausea and severe vertigo

- Proximal Occlusion (usually embolic)
  - Contralateral limb and gait ataxia
  - Contralateral hemiparesis
  - Occasional ipsilateral brainstem symptoms
- Distal Occlusion
  - Vertigo
  - Nausea

The SCA anastomoses distally with the PICA, kind of like a big bloody hug along the surface of the cerebellum. It supplies the midbrain and anterosuperior parenchyma of the cerebellum. If an isolated infarct of the SCA territory is identified on DWI imaging, the patient's care team needs to start the cardiogenic embolus workup [42] (Fig. 1.30).

5. Posterior Cerebral Artery (PCA)

A significant artery, to say the least. A PCA infarct is categorized as a posterior circulation stroke because, in most instances, the PCAs arise from the basilar terminus. However, up to 30% of people may have one or both PCAs originate from enlarged posterior communicating arteries that themselves are branches of the communicating segment (C7) of the ICA [43]. As a result, a

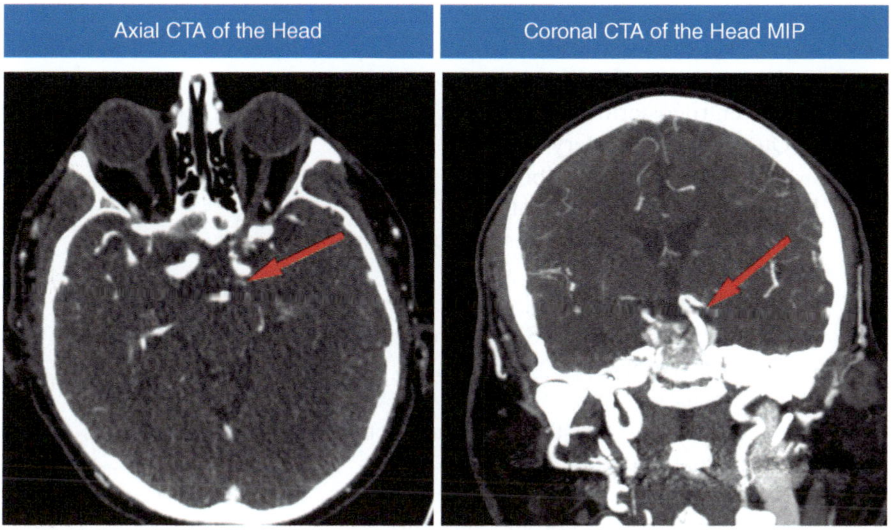

Fig. 1.30 Axial and coronal CT angiogram images annotate (red arrows) stenosis of the left P1/P2 junction PCA. Remember, when you see stenosis or occlusion of a proximal PCA segment, the midbrain and thalami are at risk

larger volume of the patient's posterior and inferior supratentorial brain parenchyma is reliant upon the anterior circulation.

- P1 Segment Pre-Communicating)
  - Interpeduncular Perforators
    - Paramedian midbrain
      - Red nucleus
      - Medial crus cerebri
  - Thalamoperforators
    - Paramedian midbrain
    - Posterior thalamus
    - Third ventricle
    - Oculomotor nucleus
    - Posterior segment of the internal capsule
  - Artery of Percheron (if you are unlucky)
    - Bilateral thalami
    - Sometimes midbrain
- P2 Segment (Post-Communicating)
  - Posterior Choroidal Arteries
    - LGN of the thalamus
    - Pulvinar/posterior thalamus
    - Choroid of lateral and third ventricles

    - Substantia nigra
    - Lateral crus cerebri
  - Thalamogeniculate Artery
    - Ventroposterior thalamus
  - Quadrigeminal Artery
    - Tectum (dorsal midbrain)
- P3 Segment (Quadrigeminal)
  - Inferior Temporal Arteries/Occipitotemporal Arteries
    - Temporal lobe
      - Fusiform gyrus
- P4 Segment (cortical)/P5 Segment (Terminal)
  - Calcarine Artery
    - Cuneus
    - /\/\/\/\/\/\/\/\/\ (It's the calcarine sulcus, ok?)
    - Lingual gyrus
  - Splenial Artery
    - Splenium of the corpus callosum

Let's occlude that PCA.

- P1 Segment [I am going to lump these together for simplification]
  - Interpeduncular perforators [medial]/thalamoperforators [44]
    - Crus cerebri/red nucleus/substantia Nigra/Oculomotor Nuclei [45]
      - Weber syndrome
        (i) Ipsilateral third nerve palsy
        (ii) Contralateral hemiplegia
      - Benedikt syndrome
        (i) Ipsilateral third nerve palsy
        (ii) Contralateral hemiplegia
        (iii) Contralateral ataxia/choreoathetosis
      - Claude syndrome
        (i) Ipsilateral third nerve palsy
        (ii) Contralateral ataxia
    - Posterior thalamus
      - Hemihypesthesia
      - Hemiataxia
  - Artery of Percheron (AoP): Paramedian trunk arising from P1
    - Bilateral thalamic infarct with or without ventral midbrain [46]
      - Great mimicker of other diseases
      - Nonspecific motor, sensory, and behavioral symptoms

- DDX: toxic, metabolic, and viral encephalopathies

(i) Carbon monoxide, hypoxic ischemic encephalopathy, and rabies, among others.
(ii) Venous infarcts can hit the bilateral thalami mimicking an AoP infarct (Fig. 1.31).

- P2 Segment
  - Posterior choroidal arteries [47]
    - LGN
      - Contralateral hemianopia
    - Pulvinar
      - Thalamic aphasia
      - Hypersomnolence
    - Choroid of the lateral and third ventricles
      - Strangely, no clinical significance [48]
    - Substantia nigra
      - Parkinsonian symptoms
    - Lateral crus cerebri
      - Contralateral extremity weakness

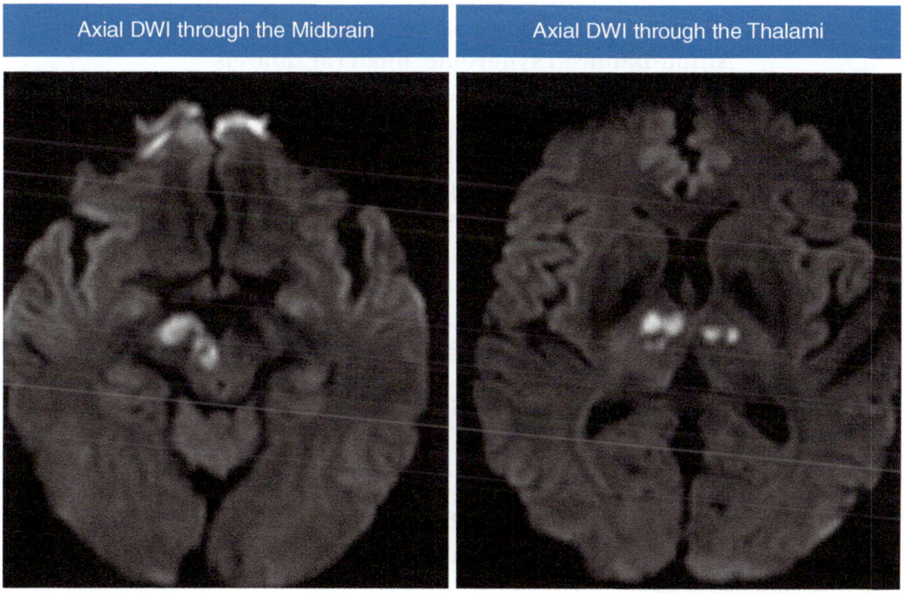

Fig. 1.31 Axial DWI images in a patient through the levels of the midbrain (left image) and thalami (right image) reveal restricted diffusion consistent with ischemic infarcts. There is involvement of the right paramedian midbrain and the right cerebral peduncle. There is additional involvement of the bilateral thalami. These findings should invoke a differential: Artery of Percheron occlusion and deep venous sinus thrombosis. In this case, it was the Artery of Percheron

- Thalamogeniculate artery
  - Ventroposterior thalamus
    - Thalamic pain syndrome (Dejerine-Roussy syndrome)

(i) Central neuropathic pain syndrome

- Quadrigeminal artery
  - Tectum (dorsal midbrain)
    - **Parinaud syndrome**
      (i) Upgaze paralysis
      (ii) Convergence retraction nystagmus
      (iii) Diplopia (fourth cranial nerve palsy)
      (iv) Colli**er's sign**

- Bilateral lid retraction on attempted upgaze. Look for lesions around the posterior commissure.
- P3 Segment
  - Anterior/temporal arteries/occipitotemporal arteries
    - Undersurface of the temporal lobe/fusiform gyrus [49]
      - Prosopagnosia: inability to recognize faces
      - Alexia: inability to read
      - Loss of dreams
- P4 Segment/P5 Segment
  - Calcarine artery: most common PCA cortical territory infarct [50]
    - Visual cortex (Brodmann area 17: cuneus and lingual gyrus)
      - Contralateral homonymous hemianopia with macular sparing
      - **Anton-Babinski syndrome**: **bilateral Infarcts**
        (i) **Cortical blindness with confabulation, which is called anosognosia in nerd-speak** [51].
    - Parieto-occipital watershed, bilateral
      - **Balint syndrome**: **often global hypoperfusion resulting in bilateral infarcts, but this syndrome has other etiologies besides stroke** [52]
        (i) Ocular apraxia: difficulty with rapid voluntary eye movements known as saccades
        (ii) Simultagnosia: inability to perceive more than one object at a time
        (iii) Optic ataxia: inaccuracy of visually-guided upper extremity movements
    - Visual association areas
      - Cerebral achromatopsia: loss of color vision
      - Akinetopsia: loss of ability to perceive motion
  - Splenial artery
    - Splenium of the corpus callosum
      - Confusion
      - Alexia without agraphia if dominant hemisphere affected

Question: An elderly gentleman and erstwhile badass (you can tell by his WW2 Veteran cap) is shouting at an elderly woman next to him, claiming he has no idea who she is. Another sad case of Alzheimer's Dementia? She tells you they've been married for 60 years and he woke up unable to recognize her or any of his war buddies' pictures. As she speaks, the patient becomes alarmed, still able to recognize her voice but unable to reconcile the disparity. An MRI with specific attention to what part of the temporal lobe is in order?

Answer: The fusiform gyrus of the right temporal lobe has been implicated in prosopagnosia.

6. Basilar Artery

   This thicc boi runs along the ventral surface of the pons in the basilar sulcus within the prepontine cistern, and it is the star player in the posterior circulation. The artery bifurcates near the pontomidbrain junction. Its branch vasculature is variable. Its territory is critical. And neurointerventionists will intervene on it with reckless abandon because the consequences of a basilar infarct can be exceptionally catastrophic and horrific.

   - Caudal Pons
     - Created by the confluence of the two vertebral arteries at the base of the pons:
       - PICA (variable origin—10% arise from the basilar artery)
         - We have discussed its supply already.
       - AICA
         - We have discussed its supply already.
     - Pontine perforators (paramedian branches/circumflex long and short)
       - Ventral (basilar) pons
         - Corticospinal/corticobulbar fibers
         - Corticopontine tracts
         - Pontine nuclei
         - Middle cerebellar peduncle pontocerebellar fibers
       - Dorsal (tegmentum) pons
         - Cranial nerve nuclei (VI, VII)
         - Paramedian pontine reticular formation (PPRF)
         - Vestibular and cochlear nuclei (VIII)
         - Spinal trigeminal nucleus
         - Reticular formation
         - Medial longitudinal fasciculus MLF
         - Medial lemniscus
         - Spinothalamic tract (more lateral)
   - Rostral Pons
     - Labyrinthine artery (occasionally from the basilar)
       - Discussed in the context of AICA already
     - Pontine perforators (paramedian branches/circumflex long and short)

- Ventral (basilar) pons
  - Pontine nuclei
  - Corticospinal/corticobulbar fibers
  - Corticopontine tracts
  - Middle cerebellar peduncle pontocerebellar fibers
- Dorsal (tegmentum) pons
  - Spinothalamic tract
  - Medial lemniscus
  - Lateral lemniscus
  - Reticular formation
  - MLF
  - Pontine reticular formation
  - Cranial nerve nuclei (V) (motor and sensory)

The basilar artery will be discussed in the context of pons strokes here, as there is a lot of ground to cover. Suffice it to say that basilar strokes are quite lethal, and if they are not lethal, they're morbid. Unlike Wallenberg syndrome, the other eponymous brainstem stroke syndromes are rare. I'll mention them in passing, but I think it's more productive to learn the brainstem geography, as the vascular supply is highly variable (Fig. 1.32).

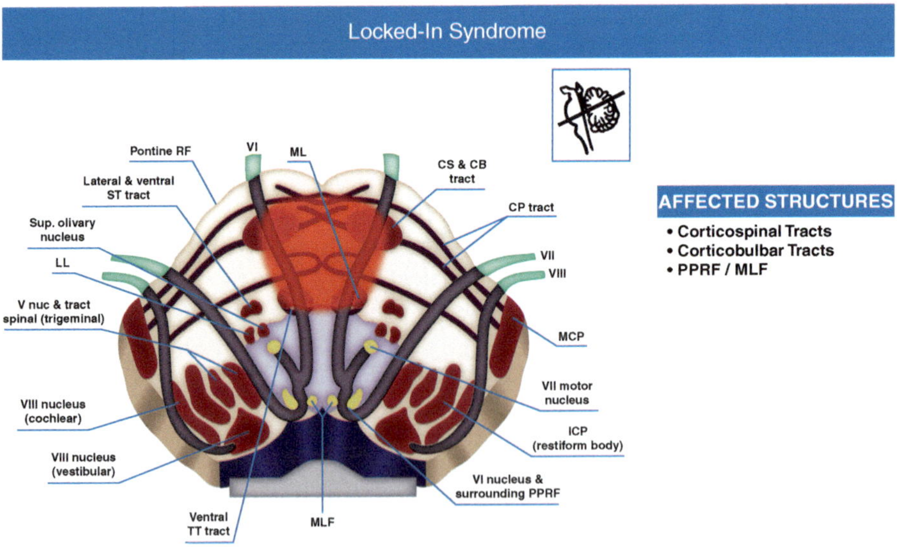

Fig. 1.32 Here we have an axial illustration through the pons. The ventral midline red zone is a generalization of the territory involved in an infarct that may cause locked-in syndrome. Be aware that every human body is different, and thus every stroke is different

- Caudal Pons [53]
  - Pontine perforators
    - Ventral (basilar) pons
      - Corticospinal/corticobulbar tracts: Note that more recent evidence suggests that the motor somatotopic organization seen in the primary motor cortex and descending pathways is not seen at lower levels. These fibers begin to mix at the level of the pons and farther caudally [54].
        (i) Contralateral lower face paralysis
        (ii) Dysarthria/dysphagia
        (iii) Contralateral body weakness
      - Pontine nuclei (corticopontine tracts)
        (i) Disruption of corticopontocerebellar pathways

(1) Pontine ataxia
      - Middle cerebellar peduncles
        (i) Purely afferent fibers
          (1) Ataxia
          (2) Hearing loss
          (3) Speech difficulty
    - Doral (tegmentum) pons
      - CN VI/VII/VIII
        (i) Ipsilateral lateral rectus palsy
        (ii) Ipsilateral facial weakness
        (iii) Vertigo, ipsilateral deafness
      - PPRF
        (i) Horizontal gaze palsy
      - Spinal trigeminal nucleus (sensory)
        (i) Ipsilateral facial numbness
      - Spinothalamic/lemniscus tracts
        (i) Loss of contralateral body pain, temperature, and fine touch and proprioception
      - Reticular formation
        (i) Impaired consciousness
      - MLF
        (i) Internuclear ophthalmoplegia

Infarction of the caudal ventral midbrain and rostral ventral pons places the patient at risk of **locked-in syndrome**. With damage to the basilar pons and the ventral tegmentum (PPRF and CN VI included), voluntary movement becomes impossible. Vertical eye movement is preserved. Consciousness is preserved because the dorsal reticular formation is not affected. If you wish to understand the impact of this condition on a human being, I highly recommend *The Diving Bell and the Butterfly* by Jean-Dominique Bauby, a man stricken with locked-in syndrome who wrote about it using only eye movements (Fig. 1.33).

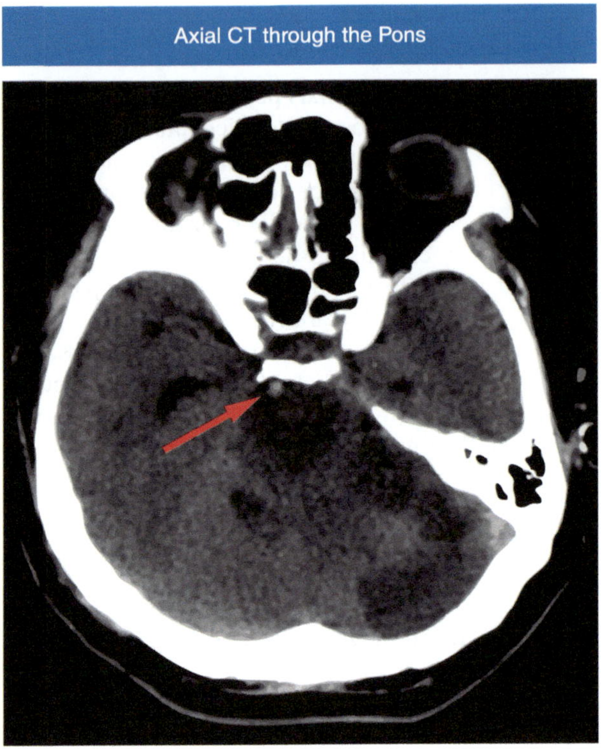

Fig. 1.33 Axial annotated noncontrast CT through the pons reveals a whole lotta badness. There is cytotoxic edema within the left cerebellar hemisphere, right paramedian cerebellum, and the pons. The red arrow points to a dense basilar artery. **This is a sign you must recognize. Failure to do so can result in patient death, locked-in syndrome, and massive lawsuits**

- Rostral Pons
  - The major differences between a rostral pons stroke and a caudal pons stroke are
    - CN V sensory and motor is more likely to be involved in a rostral pons infarct
      - Ipsilateral facial sensation loss
      - Mastication muscle weakness

During your travels, you may be confronted (assailed) by wild attendings wielding eponyms that neatly package the various stroke territories into discrete and rare syndromes. You can make a wisdom saving throw with a − 1 modifier and likely take 2d12 ego damage *or* you can use the below chart with pertinent clinical findings attached to memory devices, and roll a 20-sided die that's nothing but 20 s (Table 1.2, Figs. 1.34, 1.35, 1.36, 1.37, 1.38, 1.39, 1.40, and 1.41).

Medial midbrain syndrome/Weber
Medial midbrain syndrome/Benedikt
Medial midbrain syndrome/Claude
Dorsal midbrain syndrome/Parinaud

Table 1.2 A table of syndromes in descending anatomic position (beginning at the midbrain and proceeding caudally to the medulla) with corresponding eponymous stroke descriptions and mnemonics for each

Eponym	Brainstem location	Memory aid
Weber syndrome	Medial midbrain	These three all have CN3 oculomotor nerve palsy in common. Benedikt is weber + Claude: Benedikt has 8 letters and is the most complex
Benedikt syndrome	Medial midbrain	
Claude syndrome	Medial midbrain	
Parinaud syndrome	Dorsal midbrain	If you go to **PARIs,** you can't stand at the Eiffel tower and look up with **PARI**naud upgaze palsy (vertical gaze center)
Millard-Gubler syndrome	Ventral pons Facial-abducens hemiplegia (CN6 not originally reported)	Millard has 7 letters. Gubler has 6. Cranial nerves 7 and 6 are affected. The hemiplegia is contralateral
Foville syndrome	Ventral pons	**FOVille**. Remember FOV as field-of-view. Foville is Millard-Gubler but additional injury to PPRF (horizontal gaze center) is often cited
Marie-Foix syndrome	Lateral pons	Imagine a highly intoxicated French circus performer named *Aica*: Ataxia, nausea/vomiting, and deaf (he was shot out of a cannon, ok?)
Wallenberg syndrome	Lateral medulla	Try saying *Wallenberg* with a mouthful of peanut butter. Dysarthria, dysphagia, dysphonia. Hiccups too
Dejerine syndrome	Medial medulla	The OG (original ghost) of infarct past will kick your ass if you forget contralateral hemisensory and motor loss and ipsilateral tongue deviation!

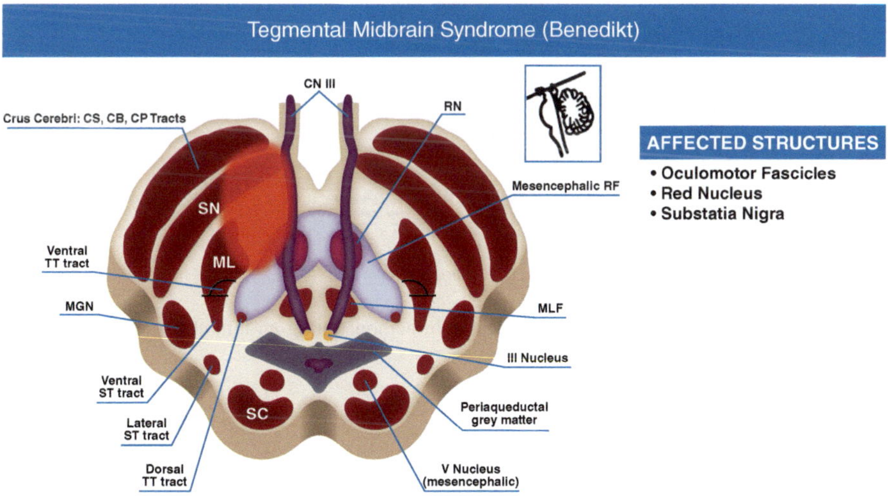

Fig. 1.34 Axial illustration through the midbrain demonstrating an infarct core (red oval) that could be responsible for Benedikt's syndrome. For the neuroradiologist, delineating the midbrain stroke syndromes is academic. For the neurologist, it's another day at the office. Seeing that "Benedikt" is longer than "Weber" and "Claude" helps me remember that Benedikt involves more structures

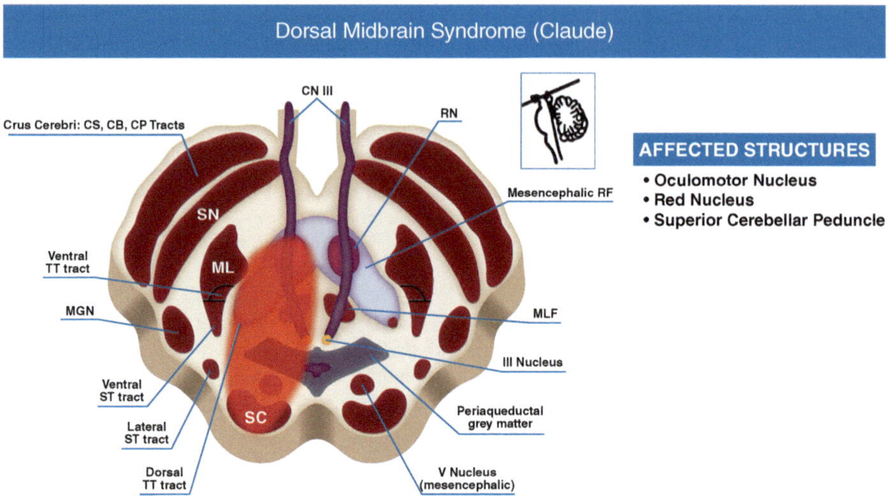

Fig. 1.35 Axial illustration through the midbrain demonstrating an infarct core (oblong red) that could be responsible for Claude's syndrome

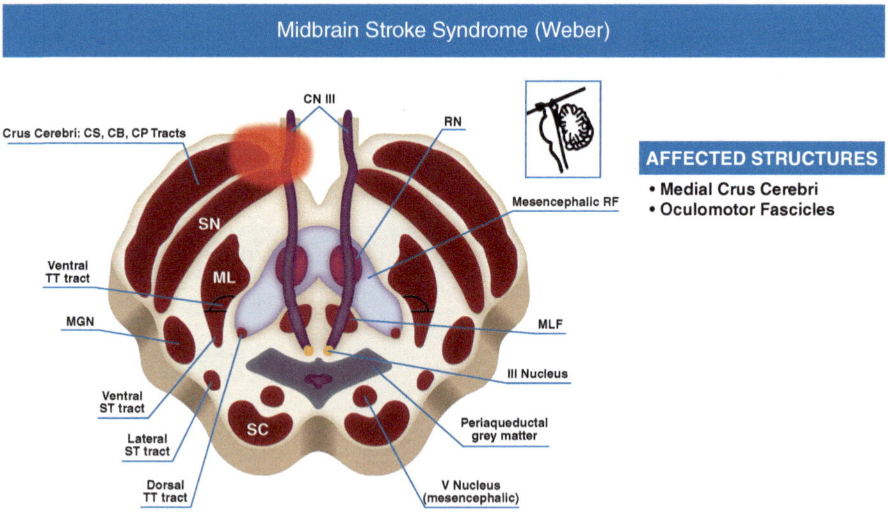

Fig. 1.36 Axial illustration through the midbrain demonstrating an infarct core (red circle) that could be responsible for Weber's syndrome

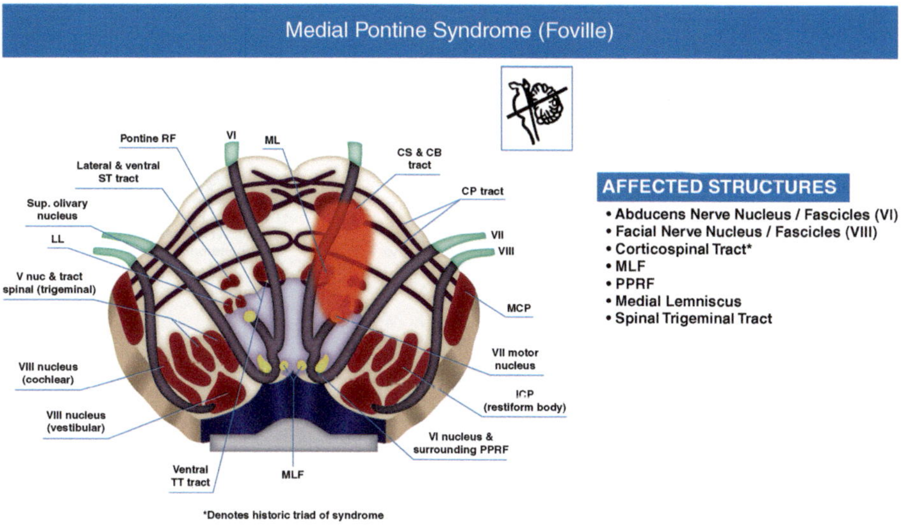

Fig. 1.37 Axial illustration through the caudal pons demonstrating a medial infarct core (oblong oval) that could result in Foville syndrome

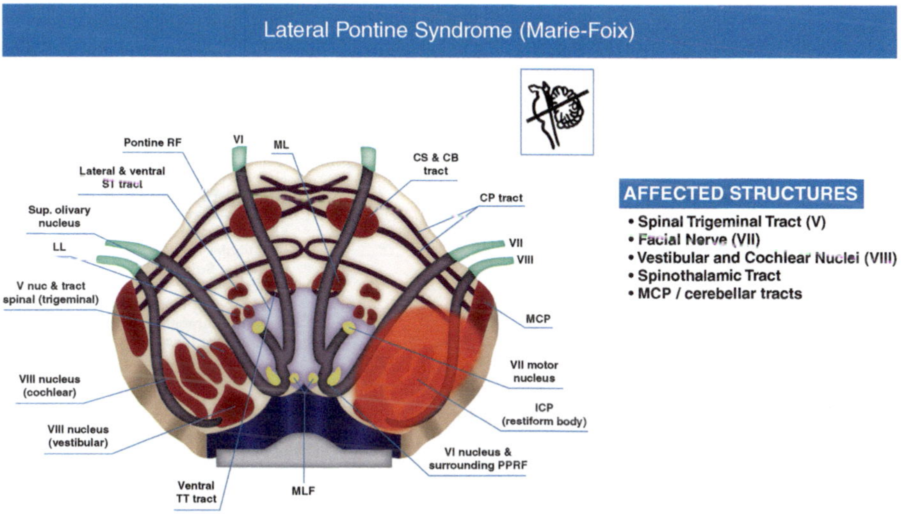

Fig. 1.38 An axial illustration through the caudal pons demonstrating an infarct core (large oval) that could be responsible for Marie-Foix Syndrome

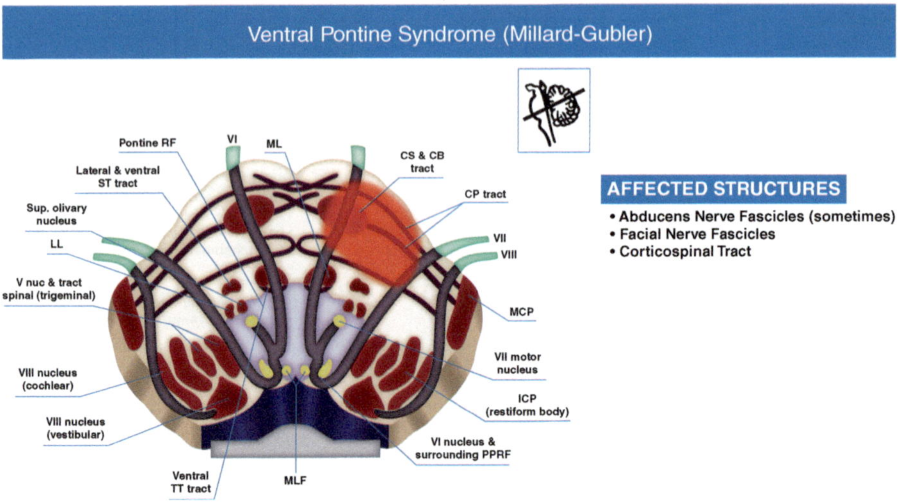

Fig. 1.39 An axial illustration through the caudal pons demonstrating a ventral infarct core (oval) that could be responsible for Millard-Gubler syndrome

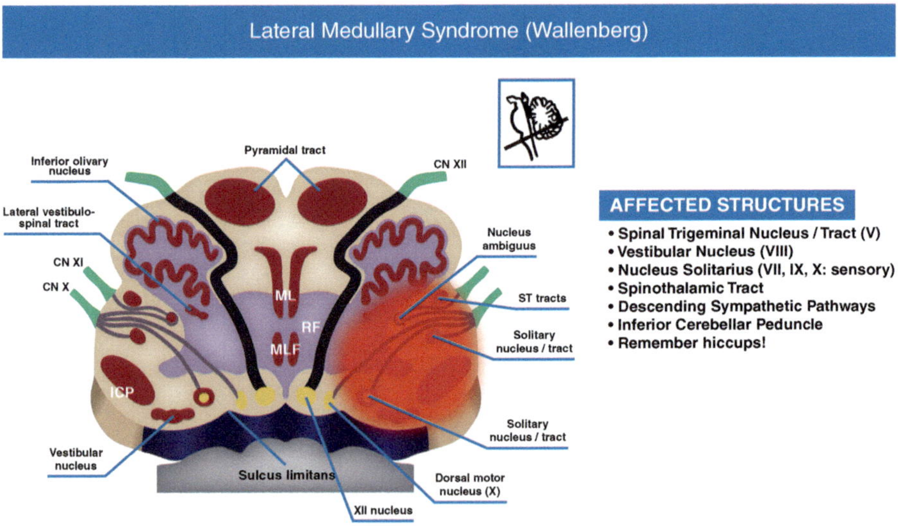

Fig. 1.40 An axial illustration through the medulla oblongata demonstrating an infarct core (oval) in the lateral medulla that could be responsible for Wallenberg syndrome

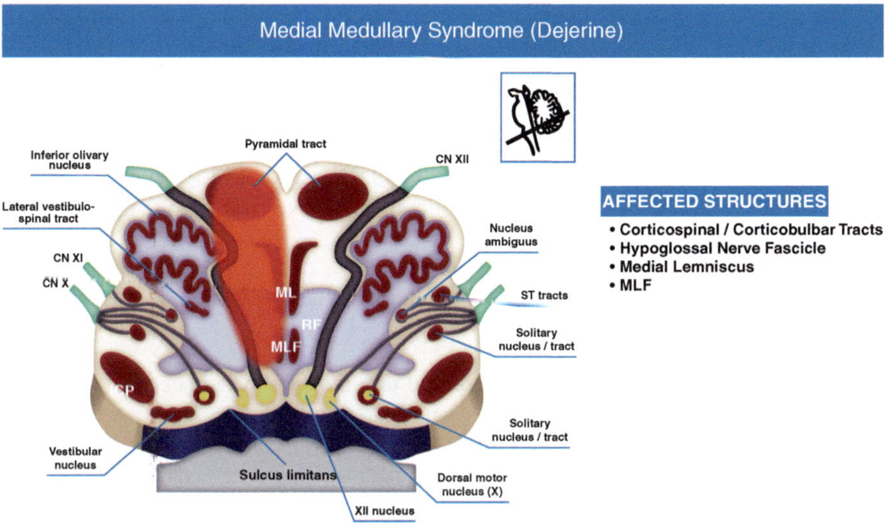

Fig. 1.41 An axial illustration through the medulla oblongata demonstrating an oblong infarct core through the medial medulla that could result in Dejerine syndrome

Ventral pontine syndrome/Millard-Gubler
Medial pontine syndrome/Foville
Lateral pontine syndrome/Marie-Foix
Lateral medullary syndrome/Wallenberg
Medial medullary syndrome/Dejerine

1.7 Perfusion Imaging

1.7.1 Bottom Line Up Front

Perfusion imaging is serial imaging of a region of the brain and the construction of a time-density curve to assess blood flow in that region. Perfusion imaging comes in two flavors: MRI and CT. It is well worth your time to learn the ins and outs of perfusion imaging because it is important for patient triage, prognostication, and intervention.

Five Diagnostic Pearls for the Resident/Fellow on Call:

- CT and MR perfusion can be used for stroke assessment but also for the evaluation of patients with masses and those who have undergone therapeutic intervention (chemoradiotherapy) on neoplasms.
- CT perfusion is much faster and less prone to artefactual corruption when compared to MR perfusion. However, MR perfusion has the advantages of better tissue contrast resolution and potentially zero contrast if arterial spin labeling is used.

- The absolute most sensitive assessment of infarct core is made by MR DWI. If an infarction is present and the tissue restricts diffusion, it's overwhelmingly dead and can be considered core. Reversal of restricted diffusion is a described entity but it is *rare*.
- The three primary variables with which you must be familiar are: mean transit time (MTT), cerebral blood volume (CBV), and cerebral blood flow (CBF).
- Perfusion imaging can be useful in delineating stroke from stroke mimics.

1. Cerebral Blood Flow Quantification Variables

   - Mean Transit Time (MTT)
     This is a measurement of the average amount of time in seconds red blood cells are within a given amount (usually 100 g) of tissue.
     - MTT is calculated by dividing cerebral blood volume (CBV) by cerebral blood flow (CBF): **MTT = CBV/CBF**.
     - Viewing MTT color maps is a helpful way of gathering a generalized picture of blood flow and visualizing penumbra.
   - Cerebral Blood Flow (CBF)
     This is a measurement of the amount of blood flowing through 100 grams of brain tissue per minute.
     - Normal values vary significantly, but values less than 10 are consistent with infarct core.
     - CBF alone is not sufficient to assess perfusion status.
   - Cerebral Blood Volume (CBV)
     This is a measurement of the amount of blood present in 100 grams of brain tissue.
   - Time to maximum (Tmax) is the time it takes tissue to reach maximum opacification relative to the time it takes for the arterial contrast bolus to reach the large vessels. Deconvolution of the arterial input function is the mathematical wizardry used to accomplish this and beyond the scope of this text.

   I highly recommend Dr. Sven Dekeyzer's YouTube content, Perfusion CT made easy, which is broken up into two parts. This is the clearest and most concise discussion of the topic I've found. Also, the remainder of his work is pure gold.

   Part 1: https://youtu.be/wjaRtG-OZk0?si=GtdAkapAf883mXXo

   Part 2: https://youtu.be/Z_sTTBcxiKg?si=jsgcNHbEY_NM0A4r

   Now a very important point I want to make about the above three values: *they are all relative* to normal contralateral blood flow. This is important because every person is different, and people with chronic disease must have their acute disease process assessed relative to their contralateral "normal."

2. CT Perfusion in Acute Ischemic Stroke

   Evaluation of stroke with CT perfusion is not a binary discriminator. Though the values described in the DAWN and DEFUSE 3 trials are often cited (and possibly testable material), the choice to intervene is clinical. Indeed, the decision to intervene in the EXTEND trial was based on clinical mismatches with core infarct volume thresholds that were age-based (Table 1.3). Furthermore, patients in EXTEND were treated with IV alteplase, NOT mechanical thrombectomy like the other two. However, there are many variables that go into the decision to intervene or not. The neurointerventionists I work with frequently indicate that they will intervene on large vessel occlusions (LVO)s with significant flexibility, even when presented with "nonoptimal" values.

   - Relative CBF (rCBF) < 30% compared to normal tissue will, by convention, be accepted as the ischemic core of an infarct. Some use rCBV to calculate the core, and there is some disagreement over which is more accurate [58, 59].
     - rCBF may *overestimate* core infarct volume if the occlusion is hyperacute (the tissue hasn't had time to infarct).
     - rCBF may *underestimate* core infarct volume after reperfusion or after collaterals have been recruited later in the infarct.
     - Maximum core volumes are between 70–100 mL. DEFUSE 3 used a maximum core inclusion value of 70 mL.
   - Tmax >6 seconds is, by convention, the tissue at immediate risk for calculating the penumbra/core ratio.
     - Perfusion mismatch (penumbra/core) ratio for DEFUSE 3 was 1.8. Other studies used a more generous ratio of 1.2.
   - The hypoperfusion index (HPI) is (Tmax >10 seconds)/(Tmax >6 seconds), and a lower ratio indicates better collaterals [60].
     - Good surrogate identifier of collateral circulation.
     - Shown to predict rate of infarct growth.

3. CT Perfusion in Stroke Mimics

   There are some stroke mimics we need to discuss (Figs. 1.42, 1.43, and 1.44). The variables can be a bit confusing, so I'll annotate them in a table (Table 1.4).

Table 1.3 The DAWN, DEFUSE III, and EXTEND studies are listed to your left. The mechanical thrombectomy criteria are described in subsequent columns for DAWN and DEFUSE. Remember that EXTEND was a chemical thrombolysis utilizing alteplase

Study	Core infarct volume	Penumbra volume	Penumbra/core ratio	LVO?	Time
DAWN	Variable	–	Clinical mismatch*	Yes	6–24H
DEFUSE III	Less than 70 mL	>15 mL	1.8	Yes	6–16H
EXTEND	Less than 70 mL	>10 mL of core	1.2	No	4.5–9H

DAWN and DEFUSE 3 used mechanical thrombolysis while EXTEND was chemical—alteplase [55–57]

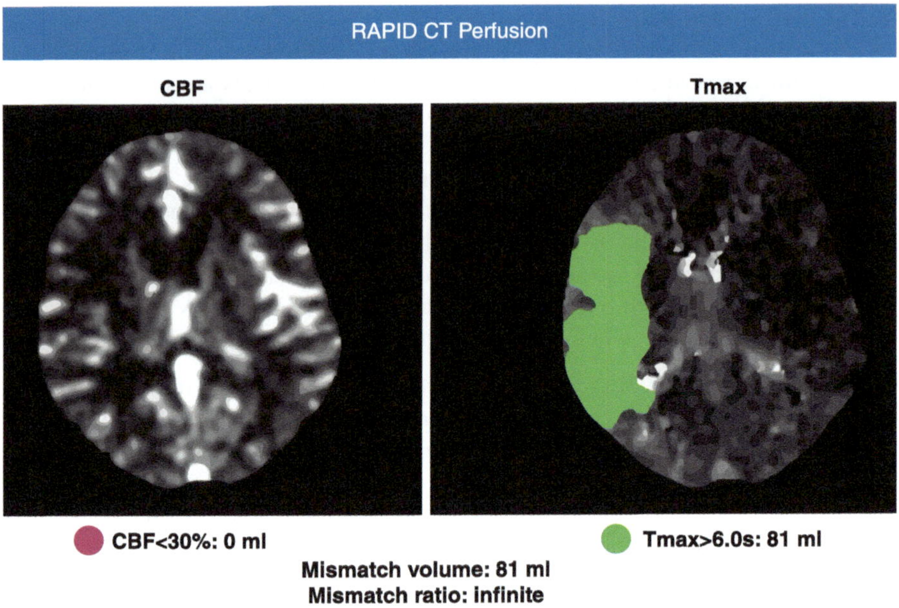

Fig. 1.42 The left image demonstrates a null color map, indicating that rCBF <30% is not detected. As a result, the infarct core is 0 mL. But is it really? On CT imaging, it's very difficult to tell. The two possibilities are: the infarct core truly does not exist because the parenchyma was saved by reperfusion/collateral circulation, **or** the infarct core does exist but was not detected. The right image demonstrates a significant penumbra as depicted in green. If the CTA demonstrates a large vessel occlusion, this patient could be a good candidate for mechanical thrombectomy

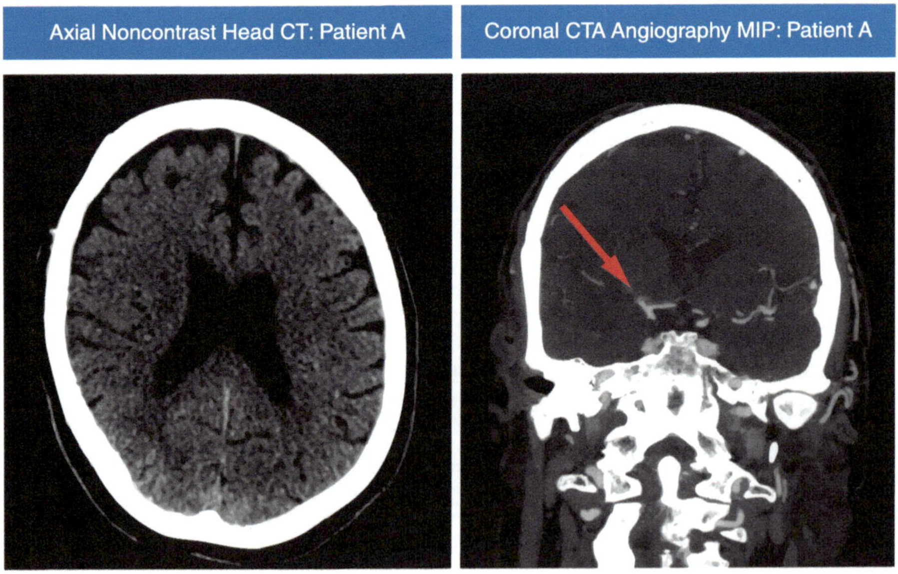

Fig. 1.43 Axial noncontrast head CT on the left reveals normal brain parenchyma. However, the coronal MIP angiogram on the right shows a proximal right M1 occlusion. Is this patient a good thrombectomy candidate? On to perfusion

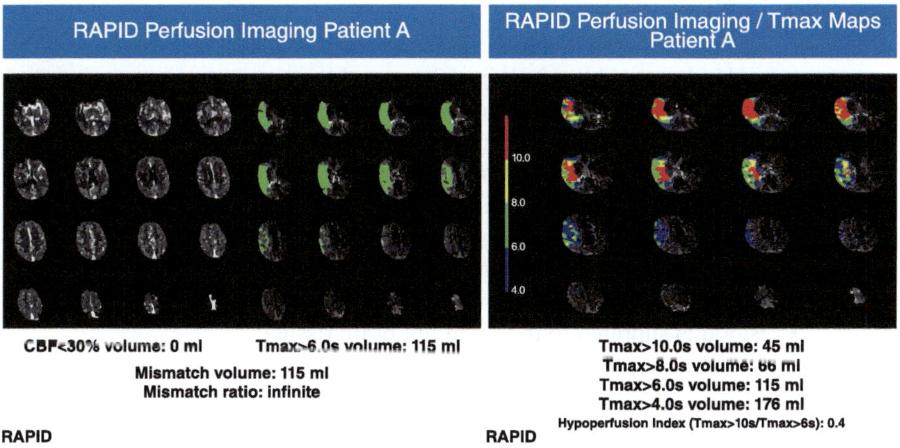

Fig. 1.44 Perfusion imaging shows no CT evidence of an infarct core because the relative CBF < 30% remains 0 mL. However, we know there is definitely a large vessel occlusion. When we review the Tmax maps, we see a penumbra volume of 115 mL. Remember, Tmax >6.0 s is the agreed-upon standard for ischemic penumbra as of 2025. Look at the red on the color map, though. The hypoperfusion index is 0.4. This is considered a low HPI, thus suggesting decent collateralization and a higher potential for a better outcome after mechanical thrombectomy

Table 1.4 Left column showing some stroke mimics. Horizontal rows representing mean transit time, relative cerebral blood flow, and relative cerebral blood volume. The arrows represent the expected finding of each value for each stroke mimic

Mimic	MTT	rCBF	rCBV
Migraine (FHM)	↑	↓	↓
Seizure	↓	↕	↕
Luxury perfusion	↓	↑	↑

- 20–50% of patients presenting to the ER for stroke symptoms will be erroneously diagnosed with stroke when in fact they have a stroke mimic [61].
  - Migraine headaches [62]
    - Hemiplegic migraine (HM)
      - Severe headaches that can be spontaneous or familial.
      - Can mimic acute infarct/Todd's paresis.
        (i) Aura phase (vasodilation)
          (1) Hypoperfusion
          (2) Decreased rCBF contralateral to side of weakness
          (3) Increased MTT
          (4) Slightly decreased rCBV
          (5) Serpiginous SWI hypointensities on MRI due to slow flow and deoxyhemoglobin paramagnetic effect
  - Seizure
    - Todd paralysis

- Poorly understood
- Presents with hemiparesis/sensory defects
- Ictal cortical hyperemia or hypoperfusion
  (i) CTP can show increased or decreased blood flow.
  (ii) Look for unusual or overlapping vascular territories.
  (iii) Assess CTA for occlusions.
  (iv) Luxury perfusion is an important differential.
- Luxury perfusion
  - Also known as postischemic hyperperfusion.
  - Thought to occur secondary to loss of autoregulation.
  - Blood is supplied to the infarcted brain tissue.
    - Increased rCBV
    - Increased rCBF
    - Decreased MTT
- CTP imaging should be correlated with noncontrast head CT. If there is an infarct, it's luxury perfusion. If not, consider Todd (Figs. 1.45 and 1.46).

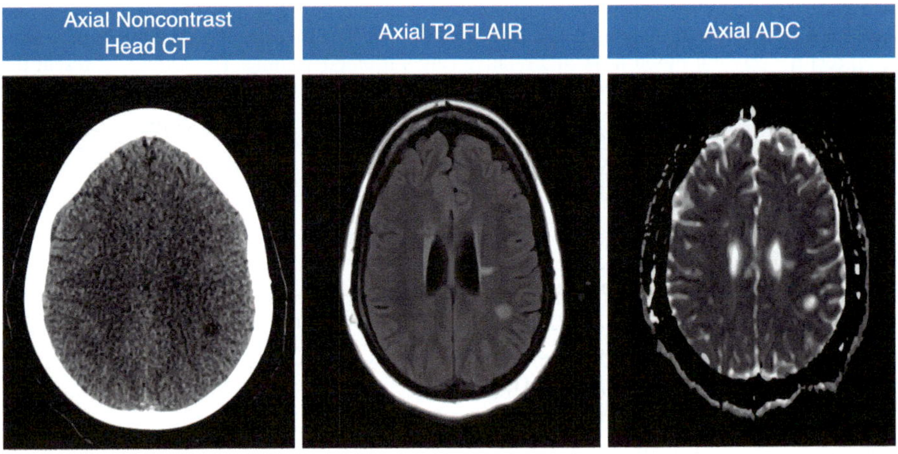

Fig. 1.45 A triad of images here tells an interesting story. A young woman presented with right-sided hemisensory loss, and a noncontrast head CT was ordered. A rounded left frontoparietal hypodensity is seen on the CT image on the left, which was interpreted as infarct. Subsequent MRI images suggest otherwise. There is no decreased diffusivity to suggest stroke on the ADC image. The middle FLAIR image shows the corresponding rounded hyperintensity and an additional oblong one that is perpendicular to the ventricle. This is multiple sclerosis masquerading as a stroke. It was the patient's first presentation with the disease, which further clouded the true diagnosis

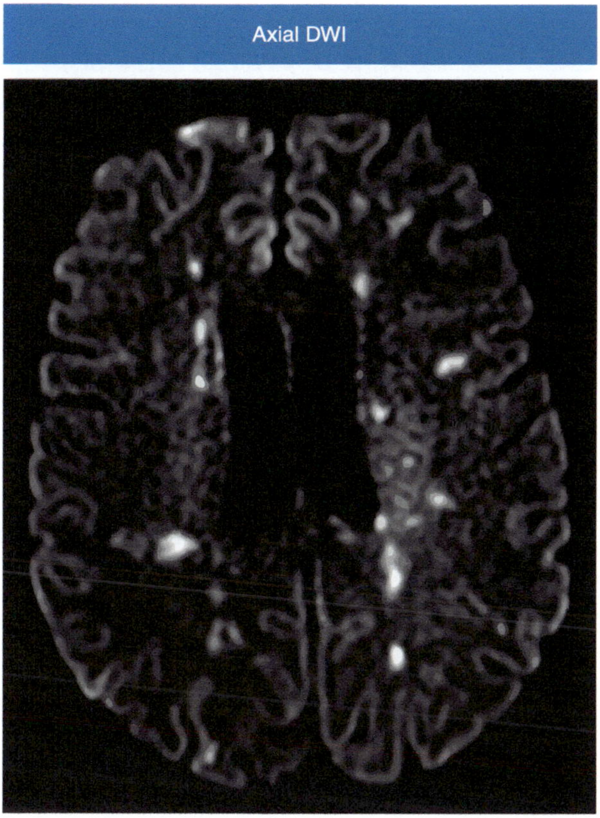

Fig. 1.46 An axial DWI image through the lateral ventricles demonstrates multiple irregular and rounded signal hyperintensities scattered through the white matter. You would not want to interpret this image in a vacuum because of the possible T2 shine-through effect. Could it be an embolic event with multiple infarcts? Yes, but you'd need to check the ADC map. Could it be multiple MS lesions? Also yes

4. Contrast Staining

   At this point, we have established some numeric values and basic clinical principles to treat a large vessel occlusion. If the patient is a good candidate, the neurointerventionists will then conduct mechanical thrombectomy, and if successful, perfusion is reestablished. In one-third or greater of patients, subsequent noncontrast head CT will show hyperdensity that overlaps with the infarct territory [63–65].

   - This phenomenon comes in two varieties: contrast staining or frank contrast extravasation.
     - Does this hyperdensity represent hemorrhagic transformation or does it represent iodinated contrast? **The distinction is clinically significant because blood would be a contraindication to further chemical thrombolysis.**

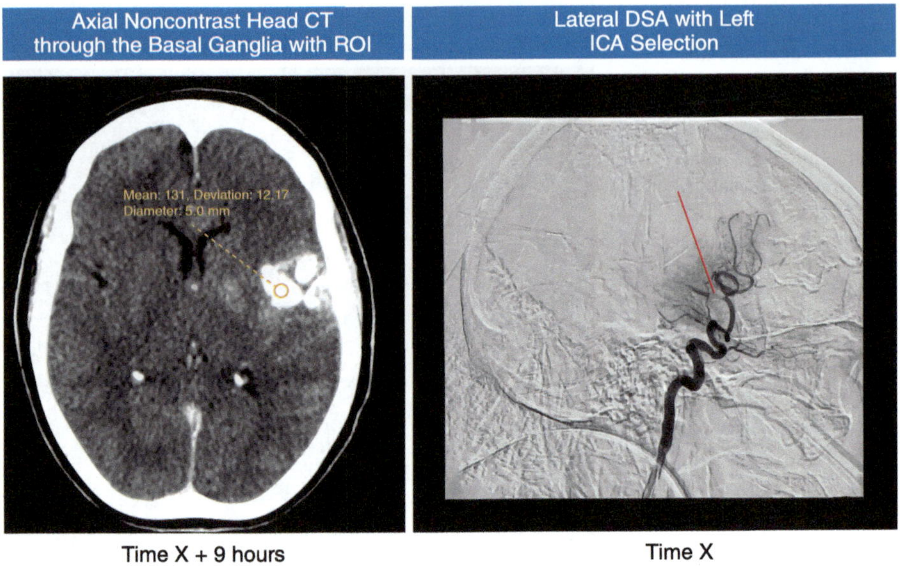

Fig. 1.47 Look first at the image on your right. This annotated lateral DSA image reveals a proximal LEFT M1 MCA occlusion for which the patient underwent mechanical thrombectomy. The image to your left reveals significant density within the left-brain parenchyma 9 hours later. An ROI placed over the largest hyperdense collection is far greater than 90 HU. It's contrast. But is all of it contrast?

- A region of interest measurement over 90HU is consistent with contrast.
- Values overlapping with blood products (30-90HU) are indeterminate on routine CT and require further imaging.
- Dual-energy CT and MRI can be useful in delineating contrast versus hemorrhage.
  - MRI is not perfect. False positives and negatives can be reported due to stagnating iodine effects on T1 and T2 relaxation times.
  - Hemorrhagic blood products are expected to remain within the brain longer than iodinated contrast. Therefore, simply waiting can be used to differentiate the two if clinically appropriate (Figs. 1.47 and 1.48).

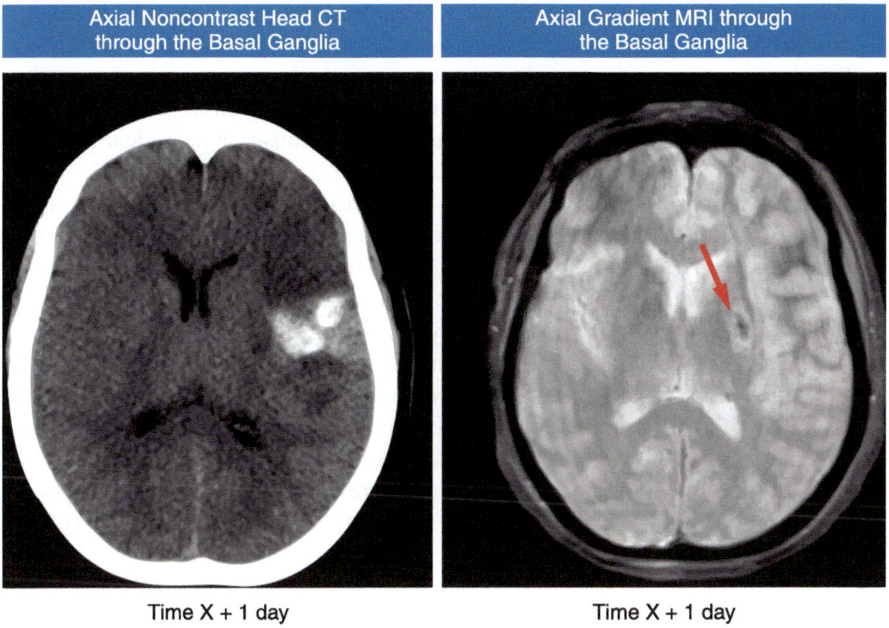

Fig. 1.48 The repeat noncontrast CT image you see on your left reveals persistent hyperdensity. Subsequent MRI on your right shows an annotated gradient image showing low signal in the left basal ganglia. The petechial microhemorrhage in the left basal ganglia does not overlap with the radiodensity seen on the CT. Therefore, one can say that most of the hyperdensity seen on CT is due to contrast

1.8 Venous Infarcts

1.8.1 Bottom Line Up Front

Venous infarcts can be protean in appearance and are definitely a minority cause of ischemic infarcts, accounting for about 3%. Regardless, they should be kept on a burner in the back of your mind, especially during summer [66], but also during viral enteritis season in the pediatric population when patients are losing excess fluids from both ends [67]. A great bit of advice I got from an attending regarding this pathology goes as follows: "if it looks weird and nonconformist, think venous infarct."

Five Diagnostic Pearls for the Resident/Fellow on Call:

- When reviewing noncontrast CTs of the head, make sure to scrutinize the dural venous sinuses. **Your suspicion for dural venous sinus thrombosis should rise if the Hounsfield Unit (HU) attenuation is around 70 or greater** [68].
- As stated above, if you can glean information or history that suggests the patient is dehydrated, increase your index of suspicion.

- Hemorrhagic transformation occurs more frequently in venous infarcts than in arterial infarcts [69].
- Seizures occur more frequently in venous infarcts than in arterial infarcts [70].
- Cortical veins can also occlude and result in venous infarcts—don't discount the diagnosis just because the large dural venous sinuses appear clear.

Honestly, a lot of the ischemic stroke discussion that occurred above is applicable to this section. Identification of the brain topography involved in the stroke is essential to correlating symptomatology. As a result, I think it would be in our best interests to review the relevant anatomy of the intracranial venous vascular system and then launch into cases for the best yield (Figs. 1.49, 1.50, 1.51, 1.52, 1.53, 1.54, 1.55, 1.56, 1.57, 1.58, 1.59, 1.60, and 1.61).

Let's have a discussion of time-of-flight imaging, shall we? Time of flight imaging is a benefit of MRI in that good angiographic images can be obtained without the use of intravenous gadolinium. Minus the extensive physics discussion, the premise of time-of-flight imaging of the vessels revolves around static suppression. You see, rapid suppression pulses can be applied to a "slab" of tissue (3D TOF) or a slice of tissue (2D TOF).

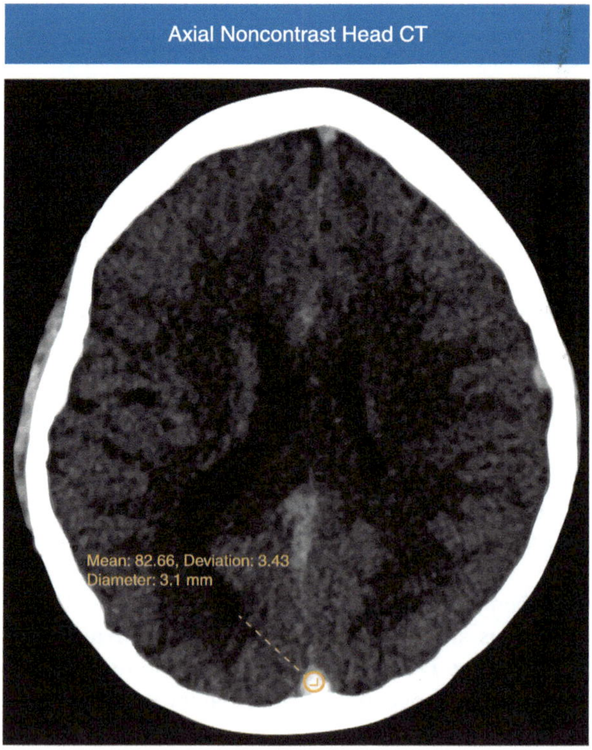

Fig. 1.49 Noncontrast CT through the head with a region of interest placed over the superior sagittal sinus, yielding a measurement of 83 Hounsfield units. As we'll discuss, a measurement greater than 70 should raise red flags. In this instance, the patient had a complete thrombosis of the superior sagittal sinus

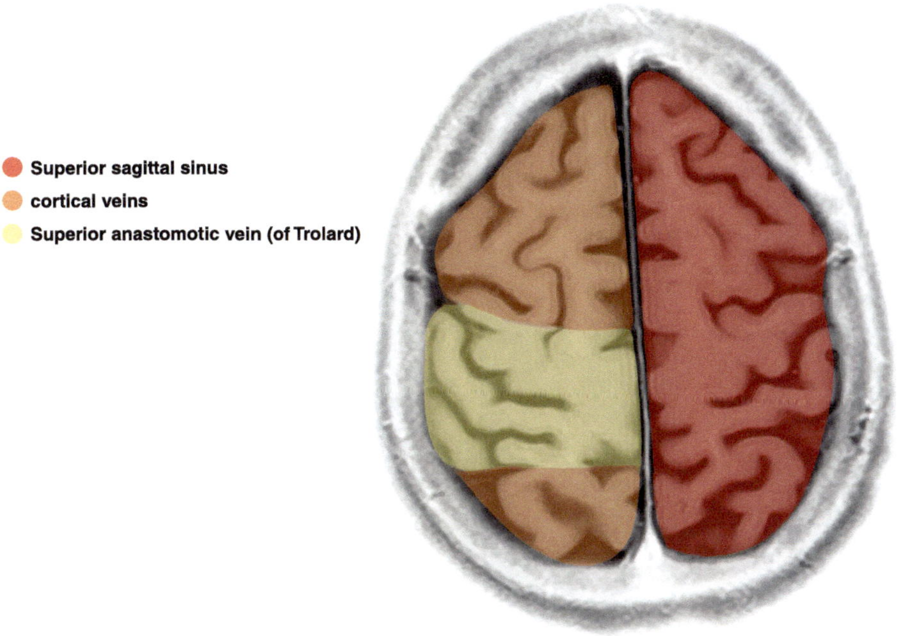

Fig. 1.50 An axial cut through the superior cortices demonstrating the venous drainage territories of the cerebral parenchyma. Note the drainage of the primary somatosensory and motor strips. (Images courtesy of Craig Hacking, Radiopaedia.org, rID: 80107 [71]. Used under license from Radiopaedia)

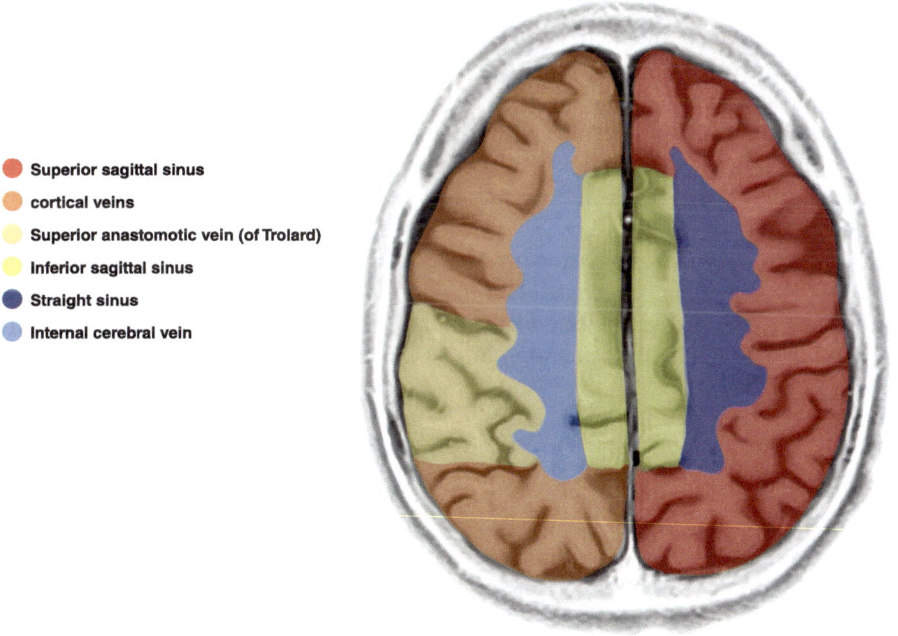

Fig. 1.51 An axial cut a bit more caudal than the first image. (Images courtesy of Craig Hacking, Radiopaedia.org, rID: 80107)

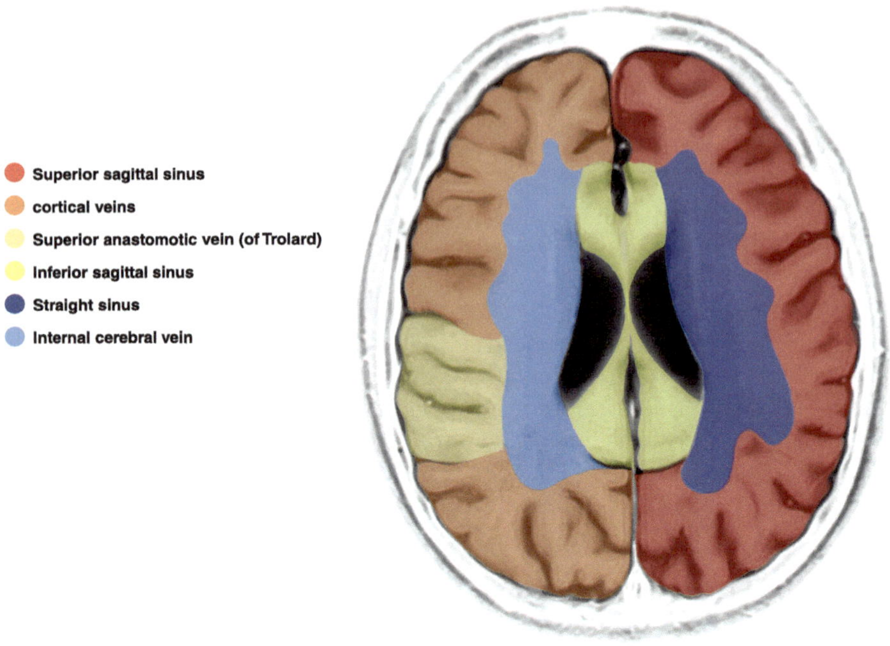

Fig. 1.52 Drainage pathways of the superior periventricular parenchyma. (Images courtesy of Craig Hacking, Radiopaedia.org, rID: 80107)

The tissue that is static, not moving, has its signal suppressed in a new steady state. This allows tissue (blood) that *is* moving to enter the field of view with normal full transverse magnetization, which appears bright due to its relatively high signal.

Saturation bands can be strategically placed to suppress arterial or venous blood, which ultimately permits MR angiography of the arteries or MR venography of the veins. Unfortunately, TOF imaging is prone to artifacts to include in-plane saturation signal loss and signal loss due to proton dephasing in regions of turbulence.

Keep in mind that rapidly flowing blood on routine T2 sequencing will present as a void (dark). This is because the rapidly flowing blood misses the 180-degree refocusing pulse. If the velocity of the blood decreases, it will be subject to both the 90-degree and 180-degree pulses and able to produce T2 signal hyperintensity.

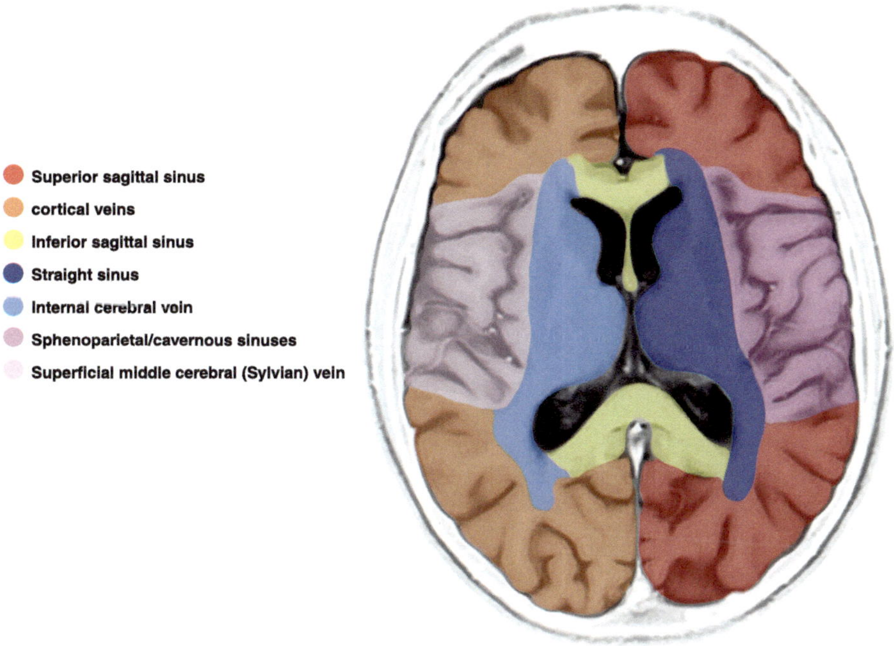

Fig. 1.53 Drainage pathways of the opercula, basal ganglia, genu and splenium of the corpus callosum, and the thalamus. Cerebral venous thrombosis of the straight sinus or vein of Galen can cause bilateral thalamic infarcts, which can mimic an artery of Percheron infarct or tip of the basilar syndrome. (Images courtesy of Craig Hacking, Radiopaedia.org, rID: 80107)

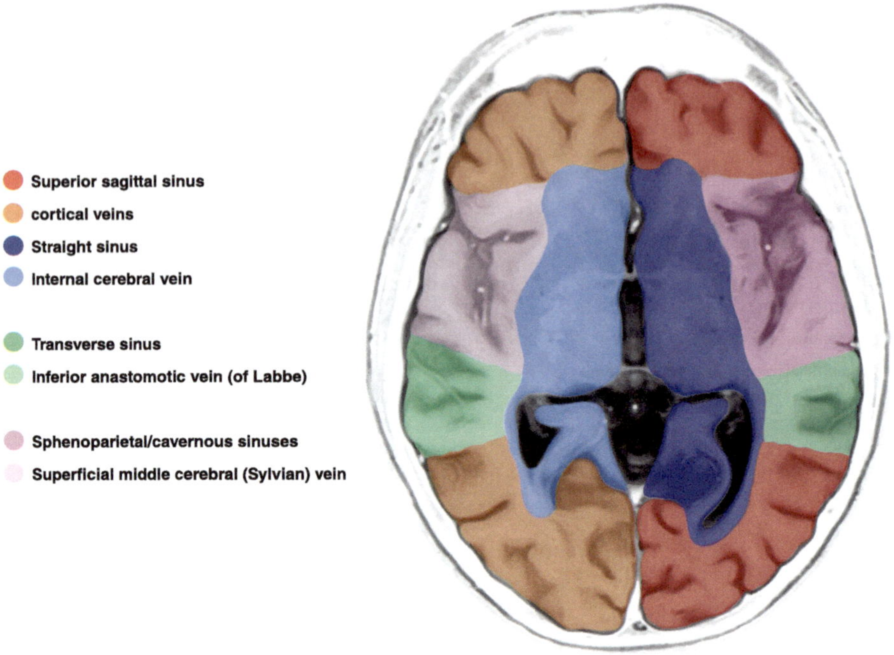

Fig. 1.54 Notice the onset of the territory drained by the vein of Labbe. Focal thrombosis of this territory has been described in the literature, and the imaging findings can be subtle. (Images courtesy of Craig Hacking, Radiopaedia.org, rID: 80107)

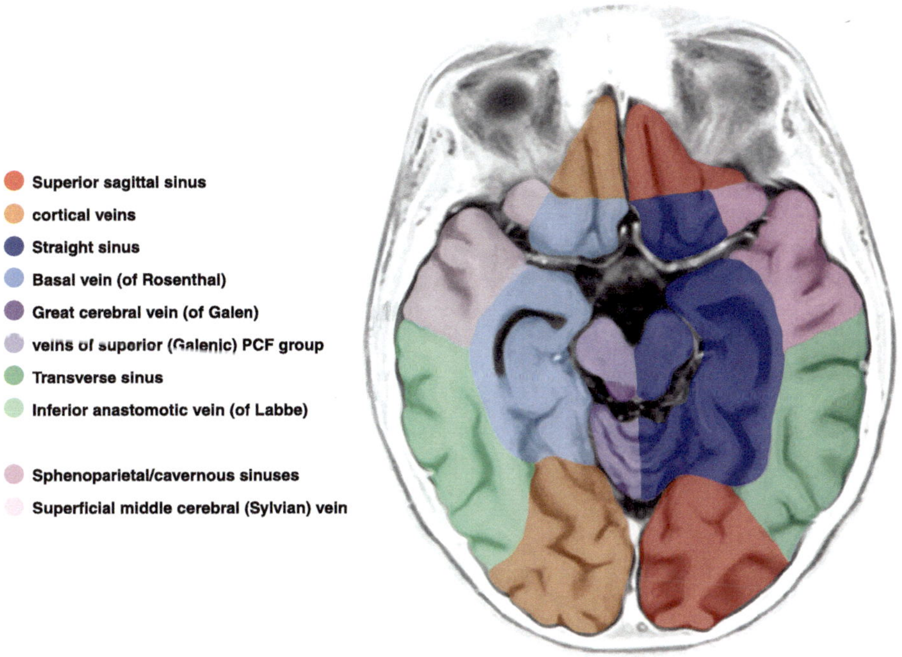

Fig. 1.55 Drainage pathways of the mesencephalic structures and superior cerebellar vermis. (Images courtesy of Craig Hacking, Radiopaedia.org, rID: 80107)

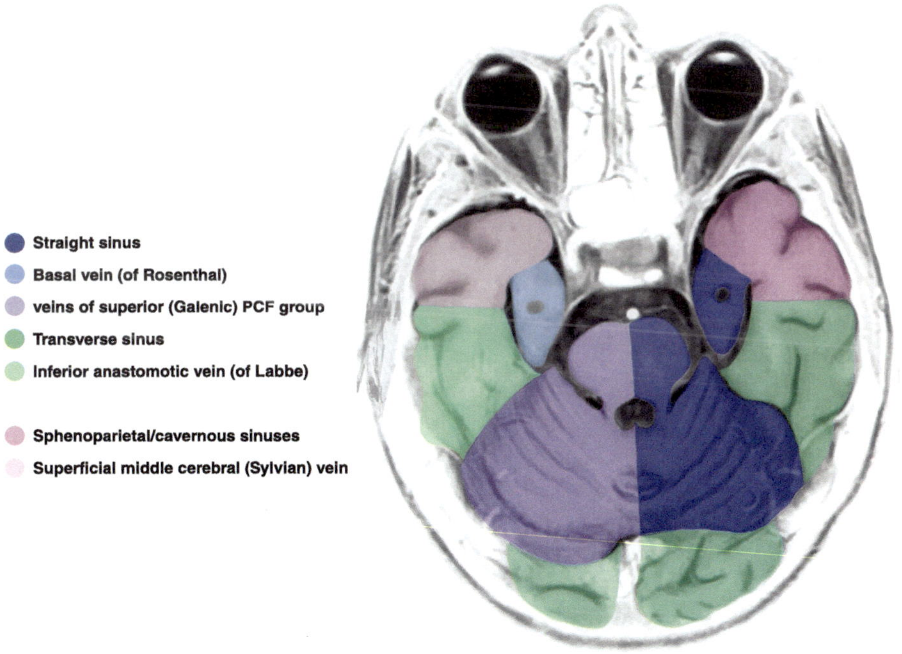

Fig. 1.56 Drainage pathways of the mesencephalic structures and cerebellum. (Images courtesy of Craig Hacking, Radiopaedia.org, rID: 80107)

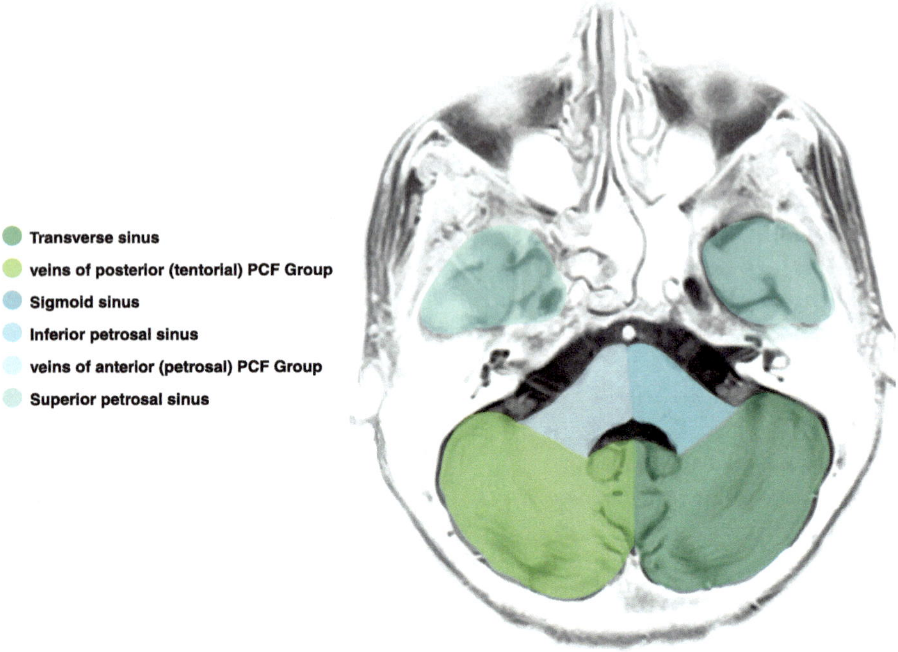

Fig. 1.57 Drainage pathways of the pons and middle cerebellar peduncles. We also catch the venous drainages of the basal surfaces of the temporal lobes. (Images courtesy of Craig Hacking, Radiopaedia.org, rID: 80107)

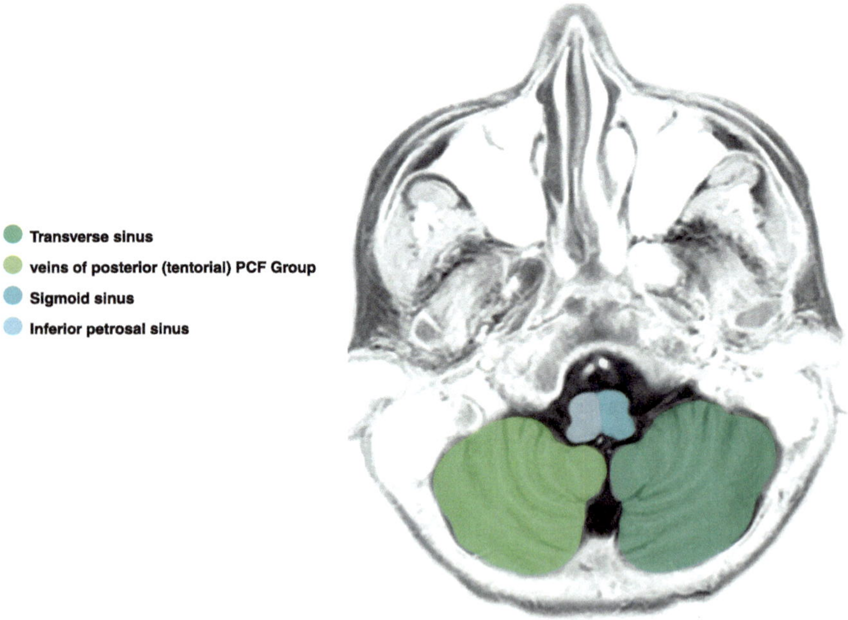

Fig. 1.58 Drainage pathways of the medulla and remainder of the cerebellum. (Images courtesy of Craig Hacking, Radiopaedia.org, rID: 80107)

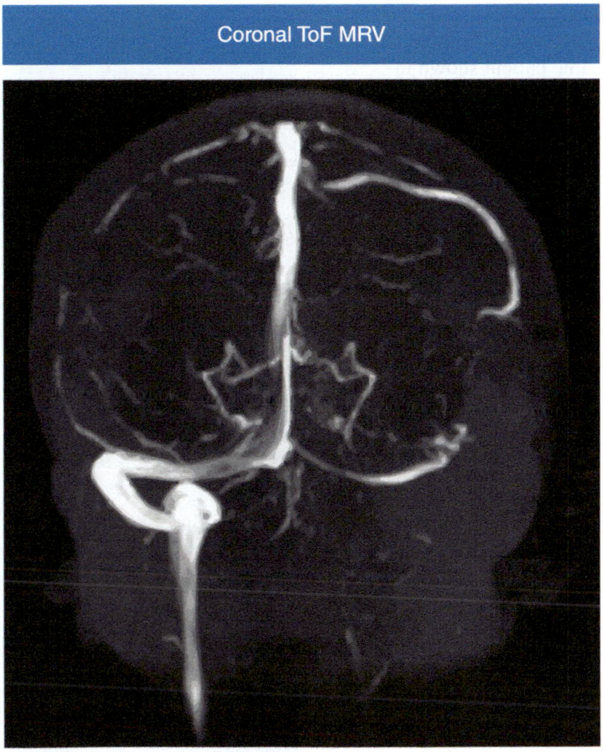

Fig. 1.59 A coronal MIP from a time-of-flight MR venogram of the head. Time of flight imaging is prone to significant artifacts and should always be correlated with routine imaging. In this case, we see a hypoplastic left transverse sinus **and** absence of flow distally in the left sigmoid sinus and jugular vein. This is abnormal

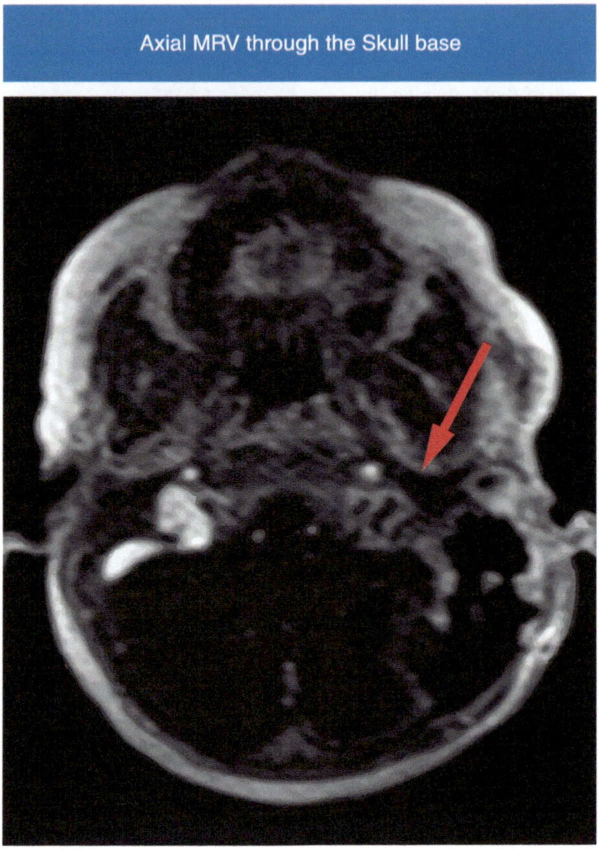

Fig. 1.60 An axial postcontrast MRV of the same patient shown in Fig. 1.43 reveals a filling defect (red arrow) in the left sigmoid/jugular junction. Posterior and lateral to this is an abscess that developed after surgical intervention

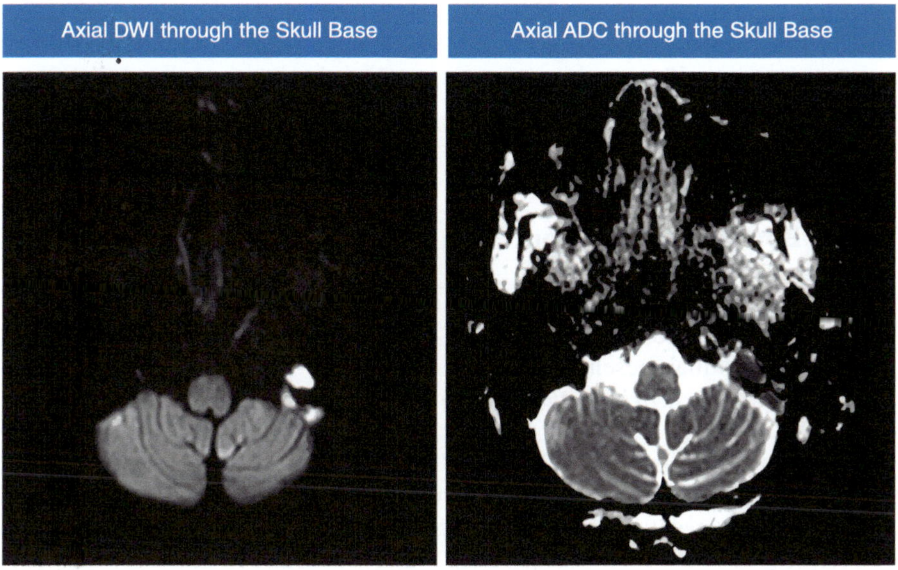

Fig. 1.61 On the left, we have an axial DWI image through the skull base with its accompanying ADC map on the right. Observe the DWI signal hyperintensity in the left sigmoid sinus and jugular vein. There is a corresponding ADC signal hypointensity seen. The diffusion restriction is real. One can say with confidence that this is a thrombus, and it is probably subacute. Without the ADC map, it could simply be T2 shine-through due to slow flow

1.9 Hemorrhagic Stroke

Hemorrhagic stroke occurs when blood leaves the confines of its vasculature through one of a multitude of pathologic processes. For the purposes of the discussion and for your diagnostic approach to intracranial hemorrhage in general, it is useful to know the patient's age and presentation. For example, one should not raise the possibility of amyloid angiopathy in a 20-year-old.

- Young (0–50)
  - Aneurysm
  - Arteriovenous Malformation
  - Cavernous Malformation
  - Reversible Cerebral Vasoconstrictive Syndrome (RCVS)
  - Posterior Reversible Encephalopathy Syndrome (PRES)
  - Spontaneous Intraparenchymal Hemorrhage (SIH)
- Old (50+)
  - Aneurysm
  - Spontaneous Intraparenchymal Hemorrhage (SIH)
  - Cerebral Amyloid Angiopathy (CAA)

- Amyloid Related Imaging Abnormalities (ARIA)
  - ARIA-H—Hemorrhage
  - ARIA-E—Effusion

1.10 Young

1.10.1 Bottom Line Up Front

The age stratification is mostly arbitrary and there is, of course, overlap in many of these disease processes. CAA is largely a disease of the elderly with the exception of some rare familial presentations.

Five Diagnostic Pearls for the Resident / Fellow on Call:

- When a noncontrast head CT with the history "worst headache of life" appears on your list, the textbook pathology to look for is subarachnoid hemorrhage. And CT has a negative predictive value of greater than 99% if performed within 6 hours [72]. Axial FLAIR sequencing is also a sensitive exam for subarachnoid hemorrhage though at least one study indicates no differences between the studies in the acute phase [73]. But lumbar puncture remains the gold standard.
- If a young patient presents with a severe headache, it is generally in the ordering provider's best interest to state whether there are mental status changes. If there are, then it is imperative that you scrutinize the temporal lobes. Herpes encephalitis is a diagnosis that cannot be missed (more on that later).
- A large volume of subarachnoid blood is not something you will miss unless you've been stricken with Anton-Babinski Syndrome while on call. After you've made the finding, take it a step further. If the blood is present largely in the basal cisterns, recommend CT angiography to look for the offending aneurysm.
- A perimesencephalic hemorrhage occurs when there is subarachnoid hemorrhage in the vicinity of the midbrain. Scant distal extension of blood is permitted, and it is thought secondary to venous hemorrhage. About 95% are spontaneous and benign. However, about 5–6% will nonetheless be the result of an aneurysm rupture [74]. Recommend the CTA and/or the digital subtraction angiography.
- Blood along the convexities is something else. It is a rarer presentation of acute subarachnoid hemorrhage. In a young person, the source could be aneurysm, but an MRI with and without contrast and a high-quality SWI sequence should be obtained if the patient is stable (Fig. 1.62).

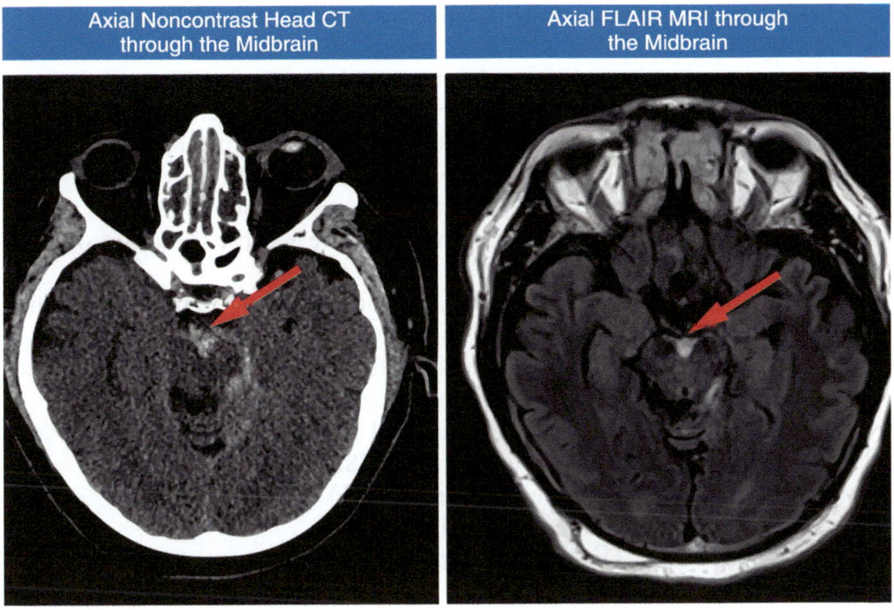

Fig. 1.62 The annotated axial CT on your left and annotated axial FLAIR on your right indicate blood in the interpeduncular cistern. There is some additional blood along the patient's left ambient cistern. When you see this, and if it is localized, raise the possibility of perimesencephalic hemorrhage. Most of these are benign, but the patient will still undergo angiography, including DSA, to prove it's not from a ruptured aneurysm

1. Aneurysm

   From the Greek *aneurysmos* meaning "dilation," these entities are an unfortunate manifestation of fluid dynamics in higher organisms with closed circulatory systems (they have been reported in animals). True aneurysms are the transmural dilation of vessel walls, often at points of hemodynamic turbulence. 90% will be saccular [75]. Pseudoaneurysms are a bit different and will be discussed in the trauma chapter.

   The prevalence of intracranial aneurysm in the general population ranges between 2 and 5% according to the literature [75–77], but is higher in individuals with family history and/or connective tissue disease. And when they rupture, about 40–50% will die [78]. Therefore, I highly, highly, *highly* recommend that you scrutinize every single brain imaging examination that comes across your desk for these little bastards. I realize this may seem like a Herculean task (and it is), but there are a few things to keep in mind that can be helpful during your career. First, aneurysms often form where blood is turbulent. This occurs at bifurcations and regions of tortuosity. Second, patients with connective tissue disorders, polycystic kidney disease, and family history of aneurysms are at higher risk. Seek this information!

- Anterior Circulation (up to 85% of intracranial aneurysms in this territory)
  - AComm
    - Comprise up to 40% of all intracranial aneurysms [79]
    - Much higher risk of rupture compared to other aneurysms [80]
  - ICA (the distal dural ring is important to know, but it cannot be seen with current imaging modalities; however, the inferior margin of the tuberculum sella is often used as a boundary in angiography [81])
    - Very important to know the anatomy here because the aneurysms have varying clinical courses:
      - Cavernous carotid artery aneurysms can produce ophthalmic symptoms.
      - If the aneurysm is oriented inferiorly and medially, carotid cave aneurysm is possible
        - (i) May have a short intradural component
          - (1) Rarely grow or rupture [82]
      - Ophthalmic segment aneurysm (5%)
        - (i) **90% are intradural**
          - (1) Often project superiorly or superomedially.
        - (ii) Remember, the ophthalmic segment of the ICA can also produce superior hypophyseal aneurysms.
          - (1) These will project medially (Figs. 1.63 and 1.64).
  - MCA Bifurcation/Trifurcation [83]
    - Account for around 18–36% of all intracranial aneurysms (Fig. 1.65)
- Posterior Circulation (Up to 15% of intracranial aneurysms in this territory)
  - PCOM [26]
    - High prevalence (~25%).

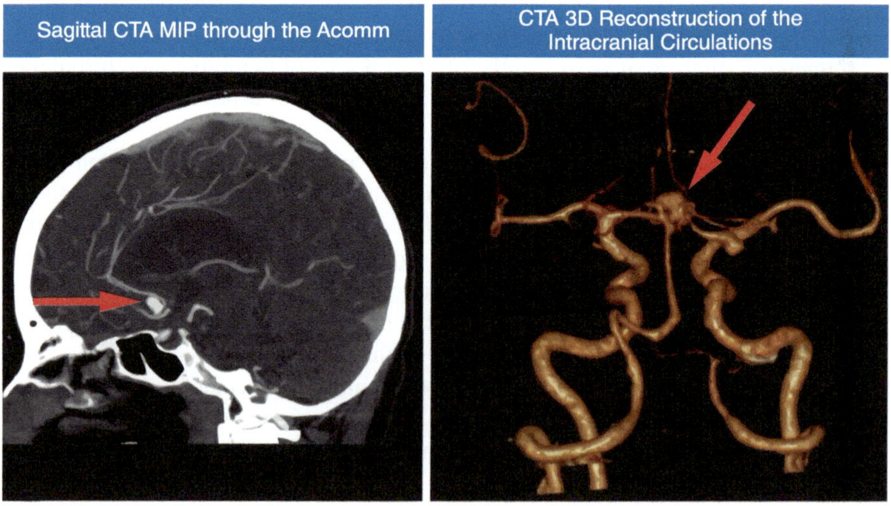

Fig. 1.63 The sagittal MIP on your left and the 3D reconstruction of the circle of Willis on your right are both annotated with arrows to draw your attention to a do-not-miss lesion: an anterior communicating artery aneurysm. The "AComm" aneurysm is common and deadly if it ruptures

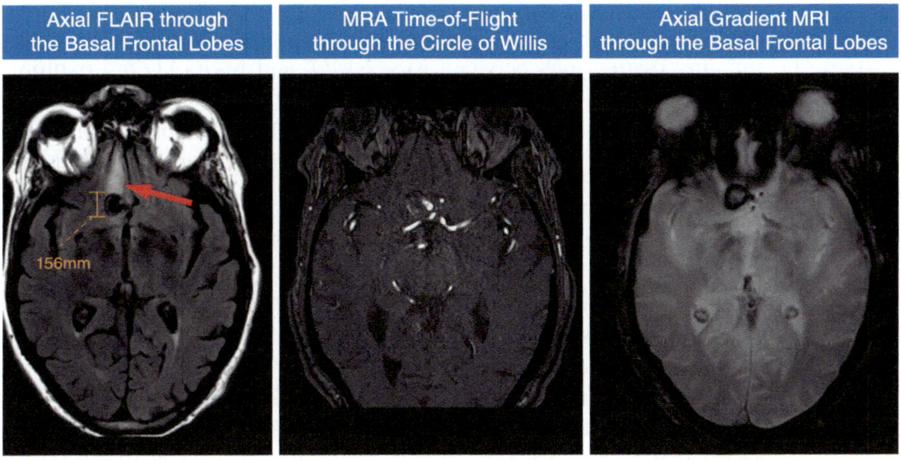

Fig. 1.64 Here we have another patient with an AComm aneurysm. The annotated axial FLAIR on your left points to a signal abnormality within the gyrus rectus. The overlapping aneurysm is well visualized both on FLAIR and gradient sequencing (right image). The MRA is less conspicuous because the aneurysm is partially thrombosed. Indeed, there are multiple case reports linking the onset of seizures to these aneurysms

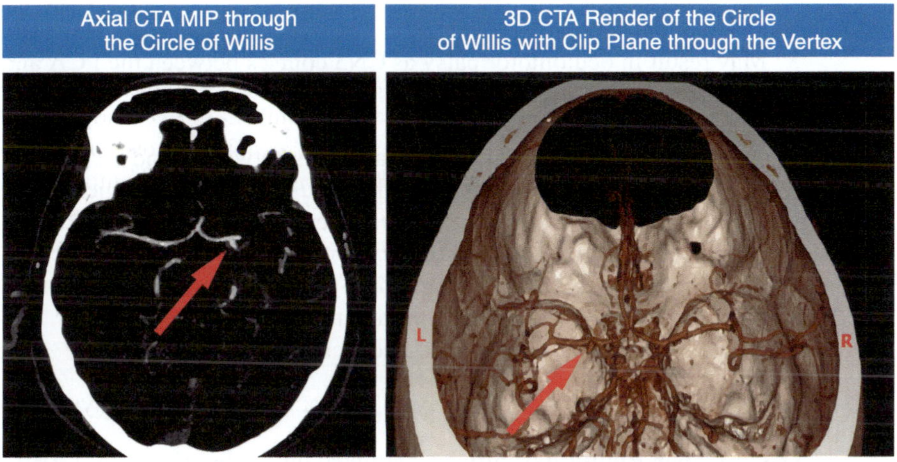

Fig. 1.65 The axial CTA image on your left with arrow demonstrates a posterior saccular outpouching at the proximal left M1 MCA. The same aneurysm is depicted on a top-down 3D reconstruction with clip plane through the vertex. Do make sure to look at your angiography imaging in all planes

    - High rate of rupture that mimics anterior circulation aneurysms.
    - May present with oculomotor nerve parasympathetic symptoms, CN 3 palsy, and eye pain.
  - SCA [84]
    - Up to 2% of aneurysms.
    - Can be subdivided into proximal and distal.
    - Prepontine/interpeduncular cistern blood.
    - Can present with CN3, 4, or 5 palsies.
  - PICA [85, 86]
    - Around 3% of aneurysms.
    - May cause brainstem compression if large.
- Closely approximates cranial nerves 9, 10, and 12.
  - May cause focal fourth ventricular hemorrhage if they rupture.
  - Basilar Tip Aneurysm [87, 88]
    - About 5–8% of aneurysms.
    - Can grow to large sizes (~25 mm).
    - Can cause blood flow abnormalities resulting in thromboembolism and stroke.
    - Can cause local mass effect including brainstem compression.
    - These big nasties are very hard to treat. Flow-diverting stents are a mainstay but these can migrate.
  - PCA Aneurysm
    - 1–2% of aneurysms.
    - May result in oculomotor palsy as CN3 courses between the PCA and SCA (Fig. 1.66).

2. Arteriovenous Malformation (AVM)

   High flow. Tangle of vessels. Nidus. Prevalence is between 5 and 10/100,000 people. Admittedly much rarer than the other causes of hemorrhagic stroke, they have a characteristic appearance, and their existence should be acknowledged as a treatable form of CNS pathology.

   AVMs have a hemorrhage risk of around 3% per annum [88]. They may exhibit their orneriness beyond simple intracranial hemorrhage, however, and are associated with seizures, mass effect, and ischemia (Fig. 1.67).

- Most commonly singular
- May be associated with syndromes
  - **Osler-Weber-Rendu/hereditary hemorrhagic telangiectasia (HHT)** has a cerebral AVM prevalence of 10% [89].
- Can be incidental to other studies or declare themselves by
  - Bleeding (most common)

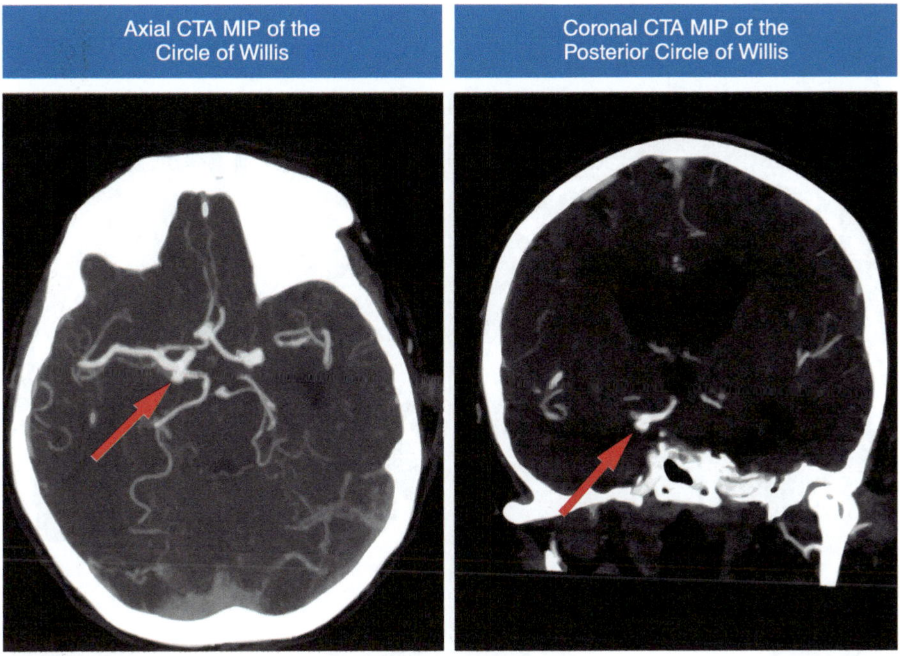

Fig. 1.66 Two annotated MIP images in the axial (your left) and coronal (your right) planes. The saccular outpouching is consistent with an aneurysm of the right fetal type PCA. This patient also has two additional aneurysms that are not annotated. Can you find them? Remember, if you see one, be alert for others

  - Seizure
  - Ischemia
- Can be graded using the ***Spetzler-Martin scale*** [90, 91]
  - Nidus Size
    - Less than 3 cm: 1 point
    - 3–6 cm: 2 points
    - Greater than 6 cm: 3 points
  - Brain Eloquence
    - Primary motor, somatosensory, visual cortex, cerebellar peduncles, internal capsule, thalamus or hypothalamus, brainstem: 1 point
    - Everywhere else: 0 points
  - Venous Drainage
    - Superficial: 0 points
    - Deep: 1 point
  - Graded 1–6 in ascending order of surgical risk with grade 6 deemed inoperable (Fig. 1.68)

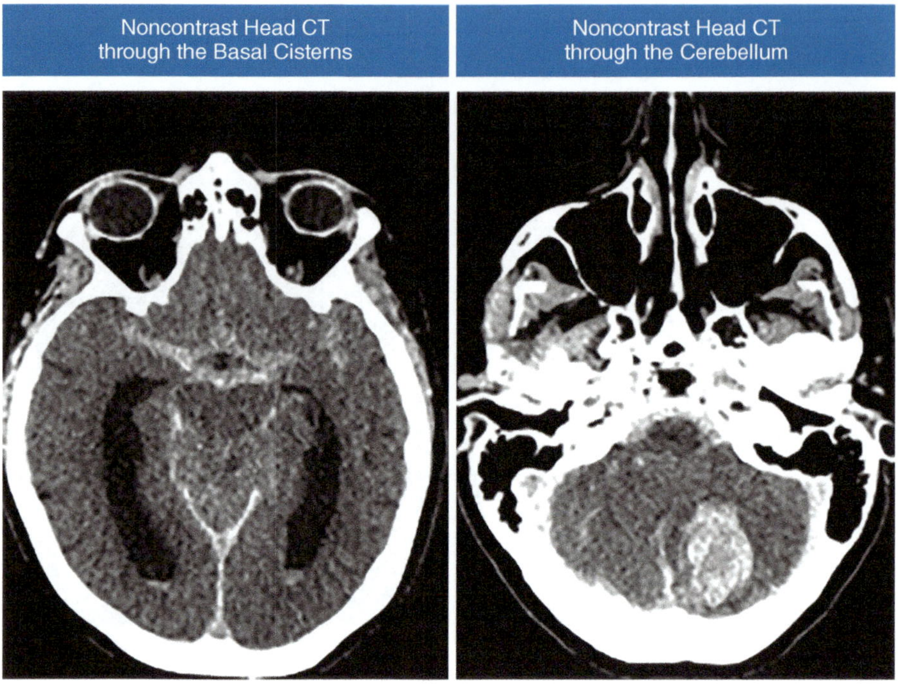

Fig. 1.67 The image on your left shows hemorrhage in the basilar cisterns, around the tentorial leaflets, and layering in the dependent ventricles. If you saw only this image in an atraumatic setting, you should consider aneurysm rupture, hypertensive hemorrhage, and AVM hemorrhage. The image on your right reveals a hemorrhage in the left cerebellum, which was proven to be an AVM on DSA

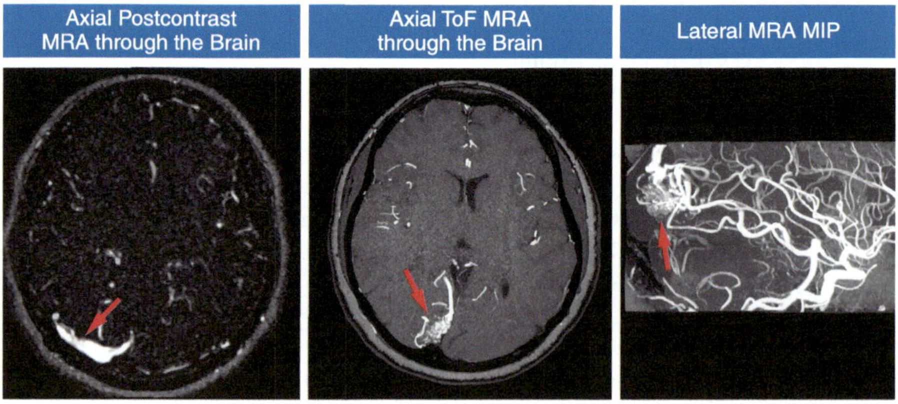

Fig. 1.68 These annotated MR images demonstrate a characteristic appearance of an AVM. The postcontrast MRA shows early venous drainage of the right transverse sinus. The middle ToF image shows the tangle of vessels as well as the enlarged right PCA feeder artery. The lateral MIP image really brings out the nidus tangle now, doesn't it?

3. Cerebral Cavernous (Cavernoma) Venous Malformation [92–94]

   Popcorn. Hemosiderin ring. Gradient hypointensity. Mulberry sign. Slow flow. These buzzwords should be ready and fast when a wily attending presents a cavernoma. These lesions can be clinically silent until their thin vessel walls rupture and cause hemorrhage. Their prevalence is about 0.50 per 100,000 persons per year.

   Cavernomas hemorrhage at a maximal rate of 11% per patient per year. They are angiographically occult and can make themselves known via mass effect or hemorrhage, which can result in focal neurologic deficits, seizure, and headache.

   - 80% are thought to be sporadic while 20% are familial.
   - The best predictor of whether a cavernoma will hemorrhage is whether it has hemorrhaged in the past.
   - If a developmental venous anomaly (DVA) is observed, a hunt for a cavernoma should commence.
   - About 7% of DVAs will have an associated cavernoma but the values vary widely. The prevalence of association increases with age.

4. Reversible Cerebral Vasoconstriction Syndrome (RCVS) [95–97]

   Thunderclap headache. This disease has a rapid onset with variable clinical symptoms. It is more common in women versus men at about 2:1. The headache is extraordinarily shocking and often occipital. It can be accompanied with migrainous symptoms.

   This disease process is more common in the young. The headache can be so rapid, intense, and severe that the patient will zoom to the ER. A noncontrast head CT will no doubt follow, which can demonstrate subarachnoid hemorrhage. It is important to note that the subarachnoid hemorrhage related to RCVS will be small volume and along the convexities. If you see blood concentrated in the basal cisterns, it's probably not RCVS.

   - RCVS is associated with pregnancy and migraine disorders.
   - It is also associated with certain medications, the most common being:
     - SSRIs
     - Sumatriptan
     - Marijuana (Maryjane!)
     - Pseudoephedrine
   - The hallmark of the disease is vasospasm that gives the blood vessels an irregular and multifocal stenotic appearance.
   - Primary angiitis of the central nervous system (PACNS) is on the differential diagnosis.
   - Severe vasospasm can result in PRES, stroke, epileptic seizures, and even Balint Syndrome (see above!).
   - RCVS is treated with calcium channel blockers.

5. Posterior Reversible Encephalopathy Syndrome (PRES) [98–101]

   Rather unfortunately named but lodged in the neuroradiology vernacular, PRES need not be posterior, won't always be reversible, and can be sans encephalopathy. But I suppose it's better than just calling it syndrome (S).

   PRES can strike anyone but middle-aged women are most often affected. Despite its name, seizures are actually the most common presenting symptom with encephalopathy or headache coming in second. The exact mechanisms are unknown but acute hypertension is often associated.

   - Increase your suspicions if the history—or self-prospected information—contains any of the following patient history:
     - Preeclampsia
     - Nephrotic syndrome/chronic kidney disease
     - Taking any of the following medications:
       - Immunomodulators
       - Bevacizumab
       - Chemotherapeutic agents: Cisplatin and carboplatin, vincristine, methotrexate
       - Glucocorticoids
   - CT of the head should come first. Assess for blood, dense basilar artery, mass, and cytotoxic edema. Look at the dural venous sinuses. In the absence of these findings, scrutinize the parenchyma for vasogenic edema.
   - If an MRI follows, assess first for restricted diffusion, hemorrhage, and dural venous sinus thrombosis. Then, look for white matter FLAIR signal hyperintensities in the parieto-occipital regions. The FLAIR abnormalities in this disease are often bilateral, but it can be unilateral.
   - Hemorrhage and microhemorrhage may be present (Fig. 1.69).

6. Spontaneous Intraparenchymal Hemorrhage (SIH) [102]

   In the young population, SIH is especially abnormal. Indeed, it can be very shocking to open a head CT on a 20-something presenting with headache or focal neurological deficits and see a glob of blood in the basal ganglia.

   Such a finding should get the *why* queries rolling in your mind as you seek an answer. Remember, if the presentation is truly spontaneous, there is almost certainly something else brewing; however, occult trauma is always a possibility. To further complicate matters, skull fracture and scalp contusion (or other traumatic findings) may follow the spontaneous hemorrhage, obfuscating the true etiology.

   - Diabetes and alcoholism are associated with spontaneous hemorrhage in young patients.
   - SIH may follow a bout of malignant hypertension.
   - Malignant hypertension may occur acutely as a result of cocaine, anabolic steroids, or amphetamine use to name a few. The use of amphetamines may be related to treatment or illicit. However, the use of nose candy is always illicit in the United States.

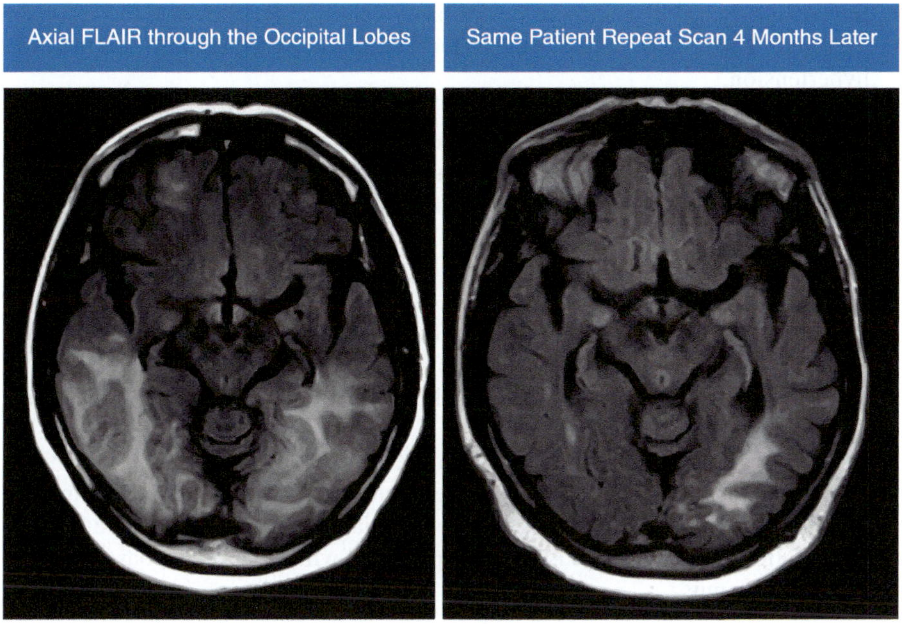

Fig. 1.69 Textbook PRES. Axial FLAIR images at different time points in the same patient. The left image demonstrates a substantial burden of white matter FLAIR signal hyperintensity, which is predominantly posterior. The right image demonstrates significant interval improvement after the offending chemotherapeutic agent was discontinued

- If malignant hypertension is present and an exogenous cause is not discovered, consider renal artery stenosis and the zebra, pheochromocytoma.
- One should also consider the possibility of hemorrhagic metastases, even in the absence of known cancer history.

1.11 Old

1.11.1 Bottom Line Up Front

As I said, the age is arbitrary. Back in medical school, I probably would have described 40 as "old," but now as I write this at the penumbra age of "silver fox" (41 years) 50 will be the delineation.

Five Diagnostic Pearls for the Resident/Fellow on Call:

- In—ahem—the more senior patient, RCVS should be much lower on your differential diagnosis (if at all) when a hemorrhagic stroke is present.
- Localize the hemorrhage if you can.

- Intraparenchymal hemorrhage in the basal ganglia is commonly the result of ruptured Charcot-Bouchard microaneurysms in the setting of chronic hypertension.
- A substantial volume of blood in the basal cisterns should precipitate a search for a ruptured aneurysm in this age group as well.
- Subarachnoid blood seen along the convexities in a more senior patient should raise the possibility of cerebral amyloid angiopathy.

1. Cerebral Amyloid Angiopathy (CAA)

   We've discussed aneurysms and SIH in the younger population. So, here we will address a new topic CAA. CAA is small and medium cortical vessel infiltration by A-Beta misfolded proteins—the same ones responsible for Alzheimer's disease.

   Pathologic analysis of the brains of deceased patients who suffered from CAA demonstrates infiltration of A-Beta into the vessel walls of medium and small cortical vessels, including the surface pial vessels. Therefore, we have an explanation as to why there is relative sparing of white matter, the basal ganglia, and the thalamus.

   - CAA is a diagnosis of exclusion.
   - When the MRI brain inevitably falls into your queue, don't neglect the gradient—or better yet—the SWI sequences.
   - If the susceptibility hypointensities are central, raise the possibility of hypertensive hemorrhages or mineralization.
   - If the susceptibility hypointensities are cortical, think CAA or perhaps small cavernomas.
   - Review the Boston Criteria if you're so inclined, which can help you categorize the findings as possible or probable (Fig. 1.70).

2. Amyloid-Related Imaging Abnormalities (ARIA)

   ARIA has come to the forefront of neuroimaging thanks to the introduction of various monoclonal antibodies purported to treat Alzheimer's disease. ARIA comes in two varieties: ARIA-H and ARIA-E, which may manifest any time during the course of taking these medications. ARIA-H is hemorrhagic and best assessed on SWI imaging, though gradient sequences are acceptable. ARIA-E reflects edema and sulcal effusion, which can be seen on FLAIR [103].

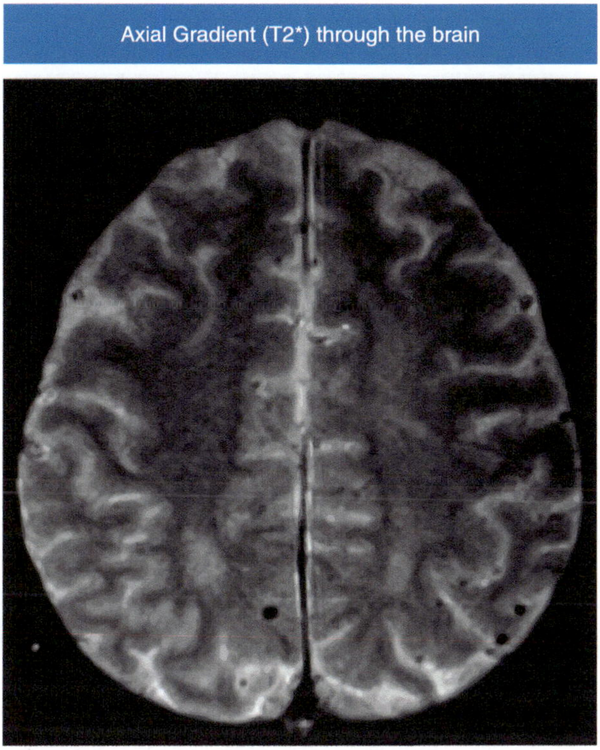

Fig. 1.70 An axial gradient signal image through the brain demonstrates multiple rounded gradient signal hypointensities. The signal abnormalities within the parenchyma are suspicious. The signal abnormalities in the sulci and extra-axial spaces may be old hemorrhage or vessels en face. In an elderly patient, such as this example, the findings are suspicious for amyloid angiopathy. If this were a 24-year-old, you'd be more concerned with prior trauma or RCVS. If this image was obtained in an elderly patient taking an amyloid targeting monoclonal antibody in the setting of Alzheimer's disease, you would call the ordering provider and raise the possibility of amyloid related imaging abnormality-hemorrhage (ARIA-H)

- ARIA-E is commonly seen, affecting about a third of individuals.
- ARIA-E is detected as subcortical FLAIR signal (edema) or sulcal effusion.
  - **This process commonly occurs in the occipital lobes and can mimic PRES.**
- ARIA-H is more important.
  - Ten or more new microhemorrhages (severe) are detected, the medication must be discontinued.
  - You must notify the provider if you discover this (Fig. 1.71).

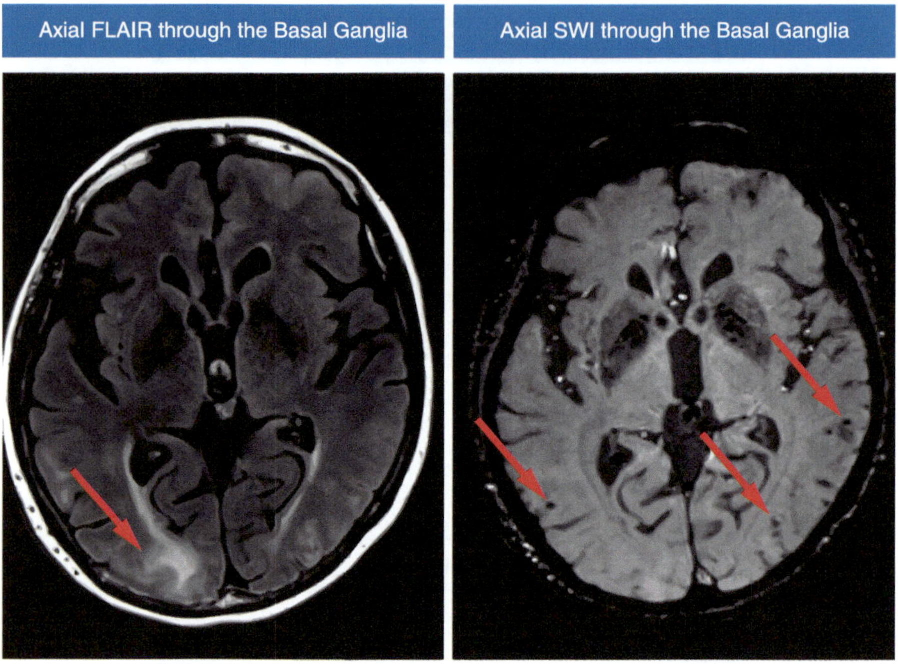

Fig. 1.71 Axial FLAIR. (left) through the basal ganglia reveals posterior paramedian subcortical FLAIR signal hyperintensity in the occipital lobe. SWI imaging through the same patient on your right shows evidence of multifocal microhemorrhages. These findings are consistent with ARIA-E and ARIA-H, and it is important that you notify the ordering provider

References

1. Ischemic Stroke. Clots. What is an ischemic stroke? American Stroke Association; 2024.
2. Saposnik G, Bushnell C, Coutinho JM, Field TS, Furie KL, Galadanci N, et al. Diagnosis and management of cerebral venous thrombosis: a scientific statement from the American Heart Association. Stroke. 2024;55(3) https://doi.org/10.1161/STR.0000000000000456.
3. Sommer P, Posekany A, Serles W, Marko M, Scharer S, Fertl E, et al. Is functional outcome different in posterior and anterior circulation stroke? Stroke. 2018;49(11):2728–32. https://doi.org/10.1161/STROKEAHA.118.021785.
4. Shapiro M, Raz E, Nossek E, Chancellor B, Ishida K, Nelson PK. Neuroanatomy of the middle cerebral artery: implications for thrombectomy. J Neurointerv Surg. 2020;12(8):768–73. https://doi.org/10.1136/neurintsurg-2019-015782.
5. Corballis M. Left brain, right brain: facts and fantasies. PLoS Biol. 2014;12:e1001767.
6. Meybodi AT, Lawton MT, Griswold D, Mokhtari P, Payman A, Benet A, et al. The anterior temporal artery: an underutilized but robust donor for revascularization of the distal middle cerebral artery. J Neurosurg. 2016;127(4) https://doi.org/10.3171/2016.8.JNS161225.
7. Kumral E, Evyapan D, Balkir K. Acute caudate vascular lesions. Stroke. 1999;30:1. https://doi.org/10.1161/01.STR.30.1.100.
8. Im I, Jun JP, Hwang S, Ko MH. Swallowing outcomes in patients with subcortical stroke associated with lesions of the caudate nucleus and insula. J Int Med Res. 2018;46(9):3552–62. https://doi.org/10.1177/0300060518775290.

9. Russmann H, Vingerhoets F, Ghika J, Maeder P, Bogousslavsky J. Acute infarction limited to the lenticular nucleus. Arch Neurol. 2003;60(3) https://doi.org/10.1001/archneur.60.3.351.
10. Emos MC, Suheb MK, Agarwal S. Neuroanatomy, Internal Capsule. 2023.
11. Navarro-Orozco D, Sanchez-Manso JC. Neuroanatomy, Middle Cerebral Artery. In:Treasure Island, FL: StatPearls. 2023. https://www.ncbi.nlm.nih.gov/books/NBK526002/
12. João RB, Filgueiras RM, Mussi ML, de Barros JEF. Transient Gerstmann syndrome as manifestation of stroke: case report and brief literature review. Dement Neuropsychol. 2017;11(2):202–5. https://doi.org/10.1590/1980-57642016dn11-020013.
13. Gammeri R, Iacono C, Ricci R, Salatino A. Unilateral spatial neglect after stroke: current insights. Neuropsychiatr Dis Treat. 2020;16 https://doi.org/10.2147/NDT.S171461.
14. Monai E, Bernocchi F, Bisio M, Bisogno AL, Salvalaggio A, Corbetta M. Multiple network disconnection in Anosognosia for hemiplegia. Front Syst Neurosci. 2020;14:14. https://doi.org/10.3389/fnsys.2020.00021.
15. Arboix A, García-Eroles L, Sellarés N, Raga A, Oliveres M, Massons J, et al. Infarction in the territory of the anterior cerebral artery: clinical study of 51 patients. BMC Neurol. 2009;9(1) https://doi.org/10.1186/1471-2377-9 30.
16. Forsberg J, Blaabjerg M, Diaz A, Gaist TA, Christensen A, Gaist D. Bilateral anterior cerebral artery infarction presenting with acute paraparesis and parkinsonism. Neurol Clin Pract. 2019;9(6) https://doi.org/10.1212/CPJ.0000000000000662.
17. Kim D-E, Park J-H, Schellingerhout D, Ryu W-S, Lee S-K, Jang MU, et al. Mapping the supratentorial cerebral arterial territories. JAMA Neurol. 2019;76(1) https://doi.org/10.1001/jamaneurol.2018.2808.
18. Arboix A, Bell Y, García-Eroles L, Massons J, Comes E, Balcells M, et al. Clinical study of 35 patients with dysarthria-clumsy hand syndrome. J Neurol Neurosurg Psychiatry. 2004;75(2) https://doi.org/10.1136/jnnp.2003.017855.
19. Hoz SS, Al-Jehani H, Aljuboori Z, BeA M, Algburi HA, Neamah AM, et al. The role of the orbitofrontal artery in the clipping of superiorly projecting anterior communicating artery aneurysms. Surgical. Neurol Int. 2021;12:12. https://doi.org/10.25259/SNI_1158_2021.
20. Alonso A, Gass A, Rossmanith C, Kern R, Griebe M, Binder J, et al. Clinical and MRI patterns of pericallosal artery infarctions: the significance of supplementary motor area lesions. J Neurol. 2011;259(5) https://doi.org/10.1007/s00415-011-6289-1.
21. Park YW, Kim CH, Kim MO, Jeong HJ, Jung HY. Alien hand syndrome in stroke – case report & neurophysiologic study. Ann Rehabil Med. 2012;36(4) https://doi.org/10.5535/arm.2012.36.4.556.
22. Tish MM, Geerling JC. The brain and the bladder: forebrain control of urinary (in)continence. Front Physiol. 2020;11:11. https://doi.org/10.3389/fphys.2020.00658.
23. Das D, Kumar R, Dutta A. Hypertrophic anterior choroidal artery: images and brief embryological overview. BMJ Case Rep. 2021;14(1) https://doi.org/10.1136/bcr-2020-240328.
24. Yu J, Xu N, Zhao Y, Yu J. Clinical importance of the anterior choroidal artery: a review of the literature. Int J Med Sci. 2018;15(4) https://doi.org/10.7150/ijms.22631.
25. Conijn MMA, Hendrikse J, Zwanenburg JJM, Takahara T, Geerlings MI, Mali WPTM, et al. Perforating arteries originating from the posterior communicating artery: a 7.0-tesla MRI study. Eur Radiol. 2009;19(12) https://doi.org/10.1007/s00330-009-1485-4.
26. Golshani K, Ferrell A, Zomorodi A, Smith TP, Britz GW. A review of the management of posterior communicating artery aneurysms in the modern era. Surgical. Neurol Int. 2010;1(1) https://doi.org/10.4103/2152-7806.74147.
27. Schmahmann JD. Vascular syndromes of the thalamus. Stroke. 2003;34:9. https://doi.org/10.1161/01.STR.0000087786.38997.9E.
28. Lineback CM, Stamm B, Rosenberg A, Bernstein RA. The case of a 70-year-old woman presenting with sudden onset apathy and amnesia. Ann Clin Transl Neurol. 2022;9(7) https://doi.org/10.1002/acn3.51575.
29. Baydin S, Baran O, Gungor A, Kuruoglu E, Tanriover N. Vascularization of the subthalamic nucleus: highlighting the significance of the Premamillary artery. World Neurosurg. 2020;135:135. https://doi.org/10.1016/j.wneu.2019.12.062.

30. Matsubara T, Suzuki K, Kawasaki A, Miyamoto M, Okamura M, Kanbayashi T, et al. Sudden onset of sleep caused by hypothalamic infarction: a case report. BMC Neurol. 2019;19(1) https://doi.org/10.1186/s12883-019-1414-3.
31. Gurley KL, Edlow JA. Avoiding misdiagnosis in patients with posterior circulation ischemia: a narrative review. Acad Emerg Med. 2019;26(11) https://doi.org/10.1111/acem.13830.
32. Gowda SN, Munakomi S, De Jesus O. Brainstem Stroke. In:Treasure Island, FL: StatPearls. 2024. https://www.ncbi.nlm.nih.gov/books/NBK560896/
33. Ali M, Amenta P, Starke R, Jabbour P, Gonzalez L, Tjoumakaris S, et al. Intracranial vertebral artery dissections: evolving perspectives. Interv Neuroradiol. 2012;18(4) https://doi.org/10.1177/159101991201800414.
34. 김 형 성 민, 김 이. 뇌신경 III–XII의 정상 자기공명영상 소견. J Korean Soc Radiol. 2020;81(3) https://doi.org/10.3348/jksr.2020.81.3.501.
35. Miao H-L, Zhang D-Y, Wang T, Jiao X-T, Jiao L-Q. Clinical importance of the posterior inferior cerebellar artery: a review of the literature. Int J Med Sci. 2020;17(18) https://doi.org/10.7150/ijms.49137.
36. Marx JJ, Thömke F, Marx JJ, Thömke F. Classical crossed brain stem syndromes: myth or reality? J Neurol. 2009;256:6. https://doi.org/10.1007/s00415-009-5037-2.
37. Park MH, Kim BJ, Koh SB, Park MK, Park KW, Lee DH. Lesional location of lateral medullary infarction presenting hiccups (singultus). J Neurol Neurosurg Psychiatry. 2005;76(1) https://doi.org/10.1136/jnnp.2004.039362.
38. Prabhu K, Woodman M. Area postrema syndrome: intractable hiccups and vomiting as a result of neuromyelitis optica spectrum disorder. JRSM Open. 2023;14(4) https://doi.org/10.1177/20542704231159601.
39. Kim JS, Han YS. Medial medullary infarction. Stroke. 2009;40:10. https://doi.org/10.1161/STROKEAHA.109.559864.
40. Lee H, Kim JS, Chung E-J, Yi H-A, Chung I-S, Lee S-R, et al. Infarction in the territory of anterior inferior cerebellar artery. Stroke. 2009;40(12) https://doi.org/10.1161/STROKEAHA.109.564682.
41. Malicki M, Szmyd BM, Bobeff EJ, Karuga FF, Piotrowski MM, Kościołek D, et al. The superior cerebellar artery: variability and clinical significance. Biomedicine. 2023;11(7) https://doi.org/10.3390/biomedicines11072009.
42. Amarenco P, Hauw JJ. Cerebellar infarction in the territory of the superior cerebellar artery. Neurology. 1990;40(9) https://doi.org/10.1212/WNL.40.9.1383.
43. Gunasekaran D, Ramakrishnan KK, Sam A, K KR, Natarajan P. Evaluating the prevalence and clinical significance of fetal posterior cerebral artery variants via magnetic resonance imaging: insights from a tertiary healthcare facility in Tamil Nadu. Cureus. 2024;16(7) https://doi.org/10.7759/cureus.64128.
44. Pai BS, Varma RG, Kulkarni RN, Nirmala S, Manjunath LC, Rakshith S. Microsurgical anatomy of the posterior circulation. Neurol India. 2007;55(1) https://doi.org/10.4103/0028-3886.30424.
45. Duncan GW, Weindling SM. Posterior cerebral artery stenosis with midbrain infarction. Stroke. 1995;26(5) https://doi.org/10.1161/01.STR.26.5.900.
46. Panart MIP, Bergua BS. Artery of Percheron in the differential diagnosis of acute altered mental status. Case Rep Med. 2021;2021 https://doi.org/10.1155/2021/5583248.
47. Neau J-P, Bogousslavsky J. The syndrome of posterior choroidal artery territory ifarction. Ann Neurol. 1996;39(6) https://doi.org/10.1002/ana.410390614.
48. Kajita M, Yanaka K, Takeda H, Saura M, Takahashi T, Aiyama H, et al. Isolated choroid plexus infarction caused by multiple occlusive cerebrovascular lesions. Radiol Case Rep. 2024;19(12) https://doi.org/10.1016/j.radcr.2024.08.075.
49. Koh YH. Right fusiform gyrus infarct with acute prosopagnosia. Acta Neurol Taiwanica. 2022;31(4):186–7.
50. Yamamoto Y, Georgiadis AL, Chang H-M. Posterior cerebral artery territory infarcts in the New England Medical Center posterior circulation registry. Arch Neurol. 1999;56(7) https://doi.org/10.1001/archneur.56.7.824.

51. Chaudhry FB, Raza S, Ahmad U. Anton's syndrome: a rare and unusual form of blindness. BMJ Case Rep. 2019;12(12) https://doi.org/10.1136/bcr-2018-228103.
52. Ribai P, Vokaer M, Tiège XD, Massat I, Slama H, Bier JC. Acute Balint's syndrome is not always caused by a stroke. Eur J Neurol. 2006;13(3) https://doi.org/10.1111/j.1468-1331.2006.01144.x.
53. Vlašković T, Brkić BG, Stević Z, Vukićević M, Đurović O, Kostić D, et al. Anatomic and MRI bases for pontine infarctions with patients presentation. J Stroke Cerebrovasc Dis. 2022;31(8) https://doi.org/10.1016/j.jstrokecerebrovasdis.2022.106613.
54. Levi AD, Schwab JM, Levi AD, Schwab JM. A critical reappraisal of corticospinal tract somatotopy and its role in traumatic cervical spinal cord syndromes. J Neurosurg Spine. 2021;36(4) https://doi.org/10.3171/2021.7.SPINE21546.
55. Albers GW, Marks MP, Kemp S, Christensen S, Tsai JP, Ortega-Gutierrez S, et al. Thrombectomy for stroke at 6 to 16 hours with selection by perfusion imaging. N Engl J Med. 2018;378(8) https://doi.org/10.1056/NEJMoa1713973.
56. Nogueira RG, Jadhav AP, Haussen DC, Bonafe A, Budzik RF, Bhuva P, et al. Thrombectomy 6 to 24 hours after stroke with a mismatch between deficit and infarct. N Engl J Med. 2018;378(1) https://doi.org/10.1056/NEJMoa1706442.
57. Ma H, Campbell BCV, Parsons MW, Churilov L, Levi CR, Hsu C, et al. Thrombolysis guided by perfusion imaging up to 9 hours after onset of stroke. N Engl J Med. 2019;380(19) https://doi.org/10.1056/NEJMoa1813046.
58. Amukotuwa S, Straka M, Aksoy D, Fischbein N, Desmond P, Albers G, et al. Cerebral blood flow predicts the infarct Core. Stroke. 2019;50(10) https://doi.org/10.1161/STROKEAHA.119.026640.
59. Otgonbaatar C, Lee JY, Jung KH, Hwang I, Yoo RE, Kang KM, Yun TJ, Choi SH, Kim JH, et al. Quantifying infarct core volume in ischemic stroke: what is the optimal threshold and parameters of computed tomography perfusion? J Stroke Cerebrovasc Dis. 2023;32(6) https://doi.org/10.1016/j.jstrokecerebrovasdis.2023.107062.
60. Guenego A, Marcellus DG, Martin BW, Christensen S, Albers GW, Lansberg MG, et al. Hypoperfusion intensity ratio is correlated with patient eligibility for Thrombectomy. Stroke. 2019;50(4) https://doi.org/10.1161/STROKEAHA.118.024134.
61. Buck BH, Akhtar N, Alrohimi A, Khan K, Shuaib A. Stroke mimics: incidence, aetiology, clinical features and treatment. Ann Med. 2021;53(1) https://doi.org/10.1080/07853890.2021.1890205.
62. Ryan DJ, Coughlan S, McSweeney N, Dineen J, Fanning N. Hemiplegic migraine as a stroke mimic: imaging and electroencephalography findings. Stroke. 2023;54(7) https://doi.org/10.1161/STROKEAHA.122.041369.
63. Whitney E, Khan YR, Alastra A, Schiraldi M, Siddiqi J. Contrast extravasation post Thrombectomy in patients with acute cerebral stroke: a review and recommendations for future studies. Cureus. 2020;12(9) https://doi.org/10.7759/cureus.10616.
64. Puntonet J, Richard M-E, Edjlali M, Hassen WB, Legrand L, Benzakoun J, et al. Imaging findings after mechanical thrombectomy in acute ischemic stroke. Stroke. 2019;50(6) https://doi.org/10.1161/STROKEAHA.118.024754.
65. Yedavalli V, Sammet S. Contrast extravasation versus hemorrhage after thrombectomy in patients with acute stroke. J Neuroimaging. 2017;27(6) https://doi.org/10.1111/jon.12446.
66. Aaron S, Lakshmanan J, Sudarsanam TD, Benjamin K, Durairaj J, Mathew V, et al. Cerebral venous thrombosis, seasonal trends, and climatic influence: a region-SpecificStudy. Ann Indian Acad Neurol. 2019;23(4) https://doi.org/10.4103/aian.AIAN_409_19.
67. Wang X-H, Zhang L-M, Chai Y-M, Wang J, Yu L-F, Zhou S-Z. Frontiers | clinical characteristics and outcomes of pediatric cerebral venous sinus thrombosis: an analysis of 30 cases in China. Front Pediatr. 2019;7 https://doi.org/10.3389/fped.2019.00364.
68. Buyck P-J, Keyzer FD, Vanneste D, Wilms G, Thijs V, Demaerel P. CT density measurement and H:H ratio are useful in diagnosing acute cerebral venous sinus thrombosis. Am J Neuroradiol. 2013;34(8) https://doi.org/10.3174/ajnr.A3469.

69. Semenov S, Moldavskaya I, Shatokhina M, Semenov A, Barbarash L, et al. How to distinguish between venous and arterial strokes and why? Neuroradiol J. 2011;24(2) https://doi.org/10.1177/197140091102400219.
70. Dias L, Carvalho M. Seizures in cerebral venous thrombosis – a retrospective analysis of a tertiary Centre cohort. Clin Neurol Neurosurg. 2023;232:232. https://doi.org/10.1016/j.clineuro.2023.107840.
71. Hacking C. Brain venous vascular territories. https://radiopaedia.org/cases/80107 (2020)
72. Blok KM, Rinkel GJE, Majoie CBLM, et al. CT within 6 hours of headache onset to rule out subarachnoid hemorrhage in nonacademic hospitals. Neurology. 2015;84(19) https://doi.org/10.1212/WNL.0000000000001562.
73. Ghaffar MKA, Eleya RZ, Nassef MA. Detection of acute small amount of subarachnoid hemorrhage: comparison between fluid-attenuated inversion recovery MR imaging and CT. Egypt J Radiol Nucl Med. 2014;45(3) https://doi.org/10.1016/j.ejrnm.2014.05.013.
74. Alrohimi A, Davison MA, Pandhi A, Abdulrazzak MA, Wadden D, Bain M, et al. Aneurysmal versus "benign" Perimesencephalic subarachnoid hemorrhage. Stroke: vascular and interventional. Neurology. 2024;4(3) https://doi.org/10.1161/SVIN.123.001166.
75. Vega C, Kwoon JV, Lavine SD. Intracranial aneurysms: current evidence and clinical practice. Am Fam Physician. 2002;66(4):601–8.
76. Johnsen LH, Herder M, Vangberg T, Kloster R, Ingebrigtsen T, Isaksen JG, Mathiesen EB. Prevalence of unruptured intracranial aneurysms: impact of different definitions – the Tromsø study. J Neurol Neurosurg Psychiatry. 2022;93(8) https://doi.org/10.1136/jnnp-2022-329270.
77. Rinkel GJE, Djibuti M, Algra A, et al. Prevalence and risk of rupture of intracranial aneurysms. Stroke. 1998;29(1) https://doi.org/10.1161/01.STR.29.1.251.
78. Aneurysm Rupture – an overview | ScienceDirect Topics. Critical Care Secrets. 2007. doi: https://doi.org/10.1016/B978-1-4160-3206-9.10068-0.
79. Pahwa B, Goyal S, Chaurasia B. Understanding anterior communicating artery aneurysms: a bibliometric analysis of top 100 most cited articles. Journal of cerebrovascular and endovascular. Neurosurgery. 2022;24(4) https://doi.org/10.7461/jcen.2022.E2022.01.001.
80. Juvela S, Poussa K, Lehto H, Porras M. Natural history of unruptured intracranial aneurysms. Stroke. 2013;44(9) https://doi.org/10.1161/STROKEAHA.113.001838.
81. Beretta F, Sepahi AN, Zuccarello M, Tomsick TA, Keller JT. Radiographic imaging of the distal Dural ring for determining the Intradural or extradural location of aneurysms. Skull Base. 2005;15(4) https://doi.org/10.1055/s-2005-918886.
82. Chandra PS. The management of incidental carotid cave aneurysms: is this really a Schrodinger's cat? Neurol India. 72(2) https://doi.org/10.4103/neurol-india.neurol-india_82_24.
83. Wang S, Li L, Gao H, Zhang K, Shao Q-J, Li T, et al. Middle cerebral artery bifurcation aneurysms are associated with patient age, sex, bifurcation angle, and vascular diameters. Sci Rep. 2023;13(1) https://doi.org/10.1038/s41598-023-50380-1.
84. Peluso JPP, et al. Superior cerebellar artery aneurysms: incidence, clinical presentation and midterm outcome of endovascular treatment. Neuroradiology. 2007;49(9) https://doi.org/10.1007/s00234-007-0251-z.
85. Barnaure I, Liberato AC, Gonzalez RG, Romero JM. Isolated intraventricular haemorrhage in adults. Br J Radiol. 2016;90(1069) https://doi.org/10.1259/bjr.20160779.
86. Peluso JP, et al. Posterior inferior cerebellar artery aneurysms: incidence, clinical presentation, and outcome of endovascular treatment. Am J Neuroradiol. 2008;29(1) https://doi.org/10.3174/ajnr.A0758.
87. Chung JY, Kang HG, Shin DS, Han HS. Rapidly growing basilar tip aneurysm causing hydrocephalus and mental deterioration. J Neurocrit Care. 2015;8(2) https://doi.org/10.18700/jnc.2015.8.2.130.
88. Theofanopoulos A, Troude L, Neyazi M, Muhammad S. Review of current treatment modalities and clinical outcome of giant saccular aneurysms of the basilar apex. Brain Spine. 2024:4. https://doi.org/10.1016/j.bas.2024.103333.

89. Brinjikji W, Iyer VN, Wood CP, Lanzino G, Brinjikji W, Iyer VN, et al. Prevalence and characteristics of brain arteriovenous malformations in hereditary hemorrhagic telangiectasia: a systematic review and meta-analysis. J Neurosurg. 2017;127(2) https://doi.org/10.3171/2016.7.JNS16847.
90. Spetzler RF, Martin NA, Spetzler RF, Martin NA. A proposed grading system for arteriovenous malformations. J Neurosurg. 1986;65(4) https://doi.org/10.3171/jns.1986.65.4.0476.
91. Kim H, Abla AA, Nelson J, McCulloch CE, Bervini D, Morgan MK, et al. Validation of the supplemented Spetzler-Martin grading system for brain arteriovenous malformations in a multicenter cohort of 1009 surgical patients. Neurosurgery. 2015;76(1) https://doi.org/10.1227/NEU.0000000000000556.
92. Kearns KN, Chen CJ, Yagmurlu K, Capek S, Buell TJ, Taylor DG, et al. Hemorrhage risk of untreated isolated cerebral cavernous malformations. World Neurosurg. 2019;131:131. https://doi.org/10.1016/j.wneu.2019.07.222.
93. Goldstein HE, Solomon RA. Epidemiology of cavernous malformations. Handb Clin Neurol. 2017:143. https://doi.org/10.1016/B978-0-444-63640-9.00023-0.
94. Meng G, Bai C, Yu T, Wu Z, Liu X, Zhang J, et al. The association between cerebral developmental venous anomaly and concomitant cavernous malformation: an observational study using magnetic resonance imaging. BMC Neurol. 2014;14(1) https://doi.org/10.1186/1471-2377-14-50.
95. Magid-Bernstein J, Omran SS, Parikh NS, Merkler AE, Navi B, Kamel H. Reversible cerebral vasoconstriction syndrome: symptoms, incidence, and resource utilization in a population-based US cohort. Neurology. 2021;97(3) https://doi.org/10.1212/WNL.0000000000012223.
96. Ly JV, Ma H, Shaloo S, Clissold B, Phan T. Convexity subarachnoid haemorrhage: a practical guide. Pract Neurol. 2023;23(5) https://doi.org/10.1136/pn-2022-003572.
97. de Boysson H, Parienti JJ, Mawet J, Arquizan C, Boulouis G, Burcin C, Naggara O, et al. Primary angiitis of the CNS and reversible cerebral vasoconstriction syndrome: a comparative study. Neurology. 2018;91(16) https://doi.org/10.1212/WNL.0000000000006367.
98. Buchhanolla P, Bir S, Angelette A, Lewis A, Kandregula S, Guthikonda B, Javalkar V. Determination of prevalence of posterior reversible encephalopathy syndrome (PRES) and its association with cerebral infarction, and outcome in the Nationwide inpatient sample, 2016–2018 (P16-10.002). Neurology. 2022;98(18_supplement) https://doi.org/10.1212/WNL.98.18_supplement.3555.
99. Parauda SC, Zhang C, Omran SS, Schweitzer AD, Murthy SB, Merkler AE, et al. Risk of stroke after posterior reversible encephalopathy syndrome. Stroke. 2022;53(11) https://doi.org/10.1161/STROKEAHA.122.038673.
100. Munirah MP, Mohamad N, Norhayati MN, Azman AN. Posterior reversible encephalopathy syndrome: a conundrum of nephrotic syndrome complication. Electronic journal of. General Med. 2023;20(2) https://doi.org/10.29333/ejgm/12835.
101. Balcerac A, Bihan K, Psimaras D, Lebrun-Vignes B, Salem J-E, Weiss N, et al. Drugs associated with posterior reversible encephalopathy syndrome, a worldwide signal detection study. J Neurol. 2022;270(2) https://doi.org/10.1007/s00415-022-11450-y.
102. Chen C-J, Brown WM, Moomaw CJ, Langefeld CD, Osborne J, Worrall BB, et al. Alcohol use and risk of intracerebral hemorrhage. Neurology. 2017;88(21) https://doi.org/10.1212/WNL.0000000000003952.
103. Agarwal A, Gupta V, Brahmbhatt P, Desai A, Vibhute P, Joseph-Mathurin N, et al. Amyloid-related imaging abnormalities in Alzheimer disease treated with anti–amyloid-β therapy. Radiographics. 2023;43(9) https://doi.org/10.1148/rg.230009.

Trauma

2

Grayson W. Hooper

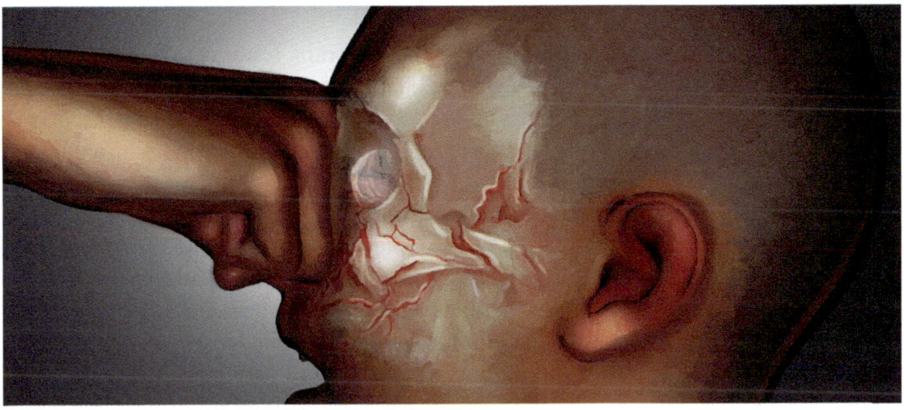

Trauma

Abstract

The trauma chapter launches into an anatomic description of the skull and dural layers followed by intracranial hemorrhage patterns. Epidural, subdural, subarachnoid, and intraparenchymal hemorrhages are reviewed in detail, including etiology, anatomy, routine case presentations, complications, and critical radiologic findings. Each is accompanied by MRI and/or CT examples. Discussion and graphic illustration of blood degradation on CT and MR imaging are interwoven into the text to help establish the chronology of hemorrhage as well as

G. W. Hooper (✉)
Department of Neuroradiology, UH/Case Western Reserve University School of Medicine, Cleveland, OH, USA

G. W. Hooper (ed.), *The Neuronomicon*,
https://doi.org/10.1007/978-3-032-05385-5_2

understand the relevant physics, which is testable material. Subsequent review of diffuse axonal injury occurs, including patient prognosis based on MRI findings. This is followed by important herniation patterns to recognize, such as uncal, cerebellar tonsillar, and subfalcine.

Craniofacial trauma review covers the anatomy of the face and calvarium. There is a review of the basic anatomy as well as the buttress system, which is useful in communicating with surgeons. Common fracture patterns, including LeFort, orbital, skull base, temporal, and calvarial are discussed with multiple examples provided. Orbital soft tissue trauma is also included.

Pediatric abusive head trauma is discussed with salient statistical, clinical, and radiologic considerations. The chapter is concluded with a discussion of spinal fractures, pseudoaneurysms and dissections, and spinal hematomas.

The human central nervous system, specifically the brain, is not very good at absorbing and redistributing force. Aside from the anatomic "crumple zones" that are the paranasal sinuses, blunt force trauma applied to the brain is like battering a scoop of gelatin. Things only get worse when a piercing mechanism is applied, like a gunshot or a crossbow bolt—*ah, Baltimore*.

Whereas stroke is a ubiquitous part of the human condition, trauma is most often very unnatural, brutal, and variable. For example, during my training in Hawaii, the most common traumatic etiology was blunt force: car accidents, scooter misadventures, surfing sorrows, and elderly patients falling out of fruit trees. In Baltimore, Iraq, and Cleveland, it was a savage combination of car accidents, gunshots, explosions, and stabbings. Consequently, your experience will vary, and so too will your competencies.

Immortan Joe Voice

Imperative 1 Satisfaction of search (or as the Europeans call it: *satiation of search*) is one of your greatest enemies in the trauma setting as a young radiologist. If you find a significant injury, annotate it and continue with your search pattern. There will often be others!

Imperative 2 *Have* a search pattern and absolutely, positively be very cautious with your templates. If you have the discipline to perform a fastidious line-item analysis based on a template, do it. However, be cautious. At best, you will look like a fool if you provide discordant information between your findings and impression sections because you left prepopulated falsities in the dictation. At worst, you can damage, delay, or even prevent proper care from being delivered.

Imperative 3 Contrast/noncontrast examination protocoling as well as bolus timing are important in the setting of trauma. Experienced technologists in tertiary care settings often have extraordinary training and knowledge; however, if you find yourself in a more rural or austere setting, you will be the person queried for specifics. Be ready. Specific discussion will follow.

2.1 Basics

We will discuss trauma from the perspective of the radiologist with respect to diagnosing injuries and sequelae that are detrimental to the life, limb, and eyesight of the patient.

- Intracranial Hemorrhage
- Craniofacial Trauma
  - Skull Fractures
  - Facial Fracture Patterns
  - Temporal Bone Fractures
  - Abusive Head Trauma
- Cervical Spine Trauma

2.2 Intracranial Hemorrhage

The brain is extraordinarily sensitive to blood, both in the context that it needs a constant blood supply to stay alive and that blood in places where it is not supposed to be can precipitate catastrophic effects. Therefore, localizing intracranial hemorrhage is of paramount importance (Fig. 2.1).

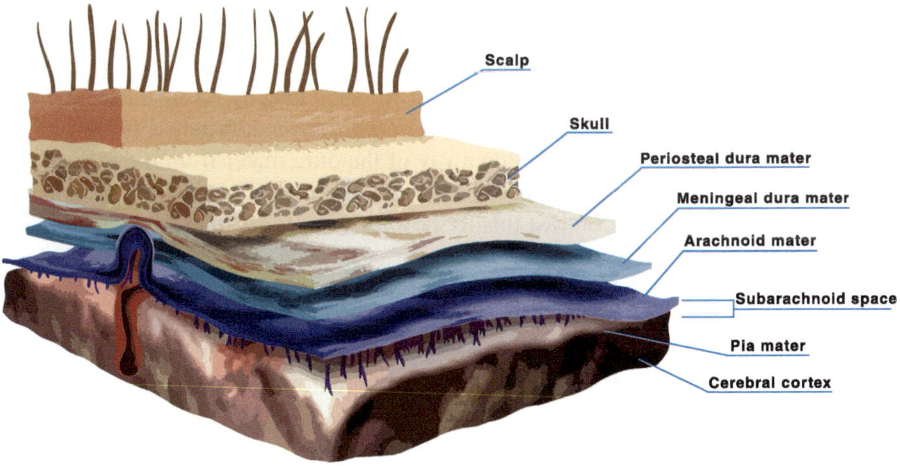

Fig. 2.1 An obliqued cross-section illustration of the layers of the dura. Though in this image there is space between the inner periosteal skull and the dura mater, this is not the case in situ, unless there is an epidural hematoma. The same goes for the potential space between the dura mater and arachnoid mater (subdural hematoma)

- Epidural Hemorrhage
  - Arterial
  - Venous
- Subdural Hemorrhage
  - Acute
  - Subacute
  - Chronic
  - Subdural Hygroma
- Subarachnoid Hemorrhage
- Intraparenchymal Hemorrhage
- Diffuse Axonal Injury
- Herniation Patterns

For the radiologist/fellow on call, intracranial hemorrhage should light a fire under your butt in the same way a stroke does, because they can grow (and kill) rapidly. However, it is here that careful analysis of the clinical history, imaging findings, and comparison studies becomes critical in proper patient disposition.

On the one hand, you want to raise the alarm *immediately* to an acute or growing hemorrhage. On the other hand, you don't want to be on the receiving end of a pissed off neurosurgeon for misinterpreting normal postsurgical changes—or chronic, stable hemorrhages for that matter—as acute emergencies. Again, trust me, I've been the very sad panda on the receiving end before. Read on!

2.3 Epidural Hemorrhage

2.3.1 Bottom Line Up Front

Epidural hemorrhage complicates 8–10% of traumatic brain injuries, and a majority of them will require surgical intervention [1]. So, what exactly is an epidural hematoma? It is a collection of blood superficial to the dura mater that is peeled away from the inner skull table. This peeling away of the dura mater from the inner skull table, often by active hemorrhage, fills the potential space with blood, creating a real space and a real problem. Contrast this to the *spinal epidural space*, which is a real and not a potential space.

Five Diagnostic Pearls for the Resident/Fellow on Call: `

- Most epidural hemorrhages (85%) are arterial and are often bound by cranial suture lines. As a result, a standard epidural hematoma will expand against the soft brain as it peels the dura mater away. It will assume a biconvex, or lentiform, shape that is limited in anterior-posterior extent [2].
- Skull fracture with arterial laceration, most often afflicting the **middle meningeal artery**, is the most common cause of epidural hemorrhage.
- Epidural hemorrhages can cross the midline, whereas subdural hematomas cannot. Furthermore, about **10% of epidural hematomas *can* cross suture lines** [3].

- Venous epidural hemorrhages are possible secondary to bleeding from dural venous sinuses or emissary veins. These can cross the midline, displace dural venous sinuses, and cause venous infarctions.
- Spontaneous epidural hematomas are possible, so do not rule them out in the absence of trauma. They can occur in the setting of coagulopathy, craniofacial infection, and sickle cell disease, among others [4, 5].

1. Arterial Epidural Hemorrhage
   The textbook history for this process is trauma to the squamous temporal bone, often from fighting, motor vehicle accidents, or baseball injuries. **Young patients are more prone to developing expansive epidural hemorrhages due to the fact that the dura mater is not as tightly adhered to the inner skull table as in adults**. Much ado is made of the so-called "lucid interval" after an injury, which is reported in up to 50% of epidural hematoma cases. In the lucid interval, the patient is initially knocked out, followed by a return to near-normalcy, followed by rapid deterioration (Figs. 2.2 and 2.3).
2. Venous Epidural Hemorrhage
   These strange fellows sometimes occur after the laceration of a dural venous sinus. They may appear quite unusual, even to the trained eye, because they are

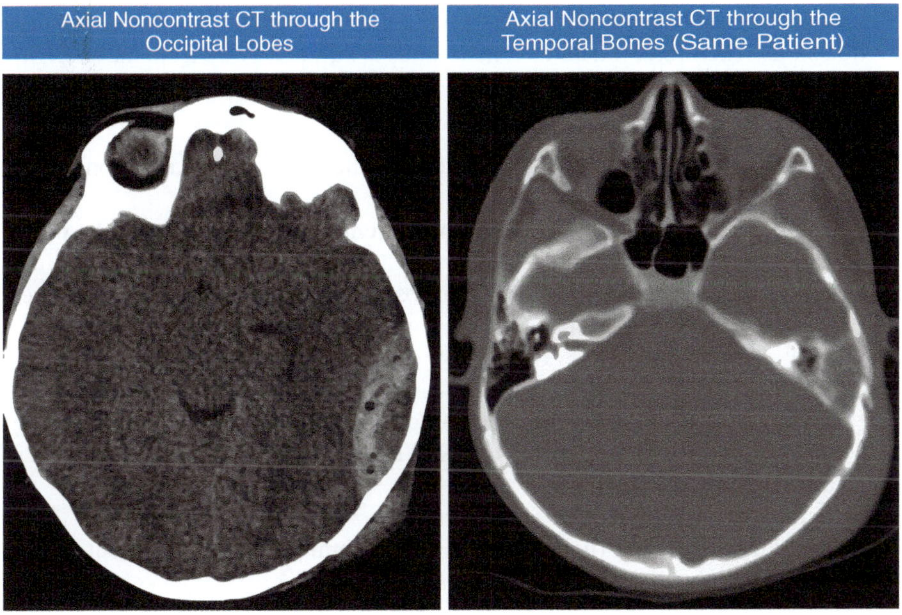

Fig. 2.2 The axial head CT image on your right windowed for air reveals asymmetric opacification of the patient's left mastoid air cells. Additional scrutiny reveals a fracture. The CT image on your left, a few slices superior, demonstrates a lentiform fluid collection of mixed density with indwelling foci of air. This finding is consistent with an epidural hematoma that has "imbibed" air after fracture through the mastoid

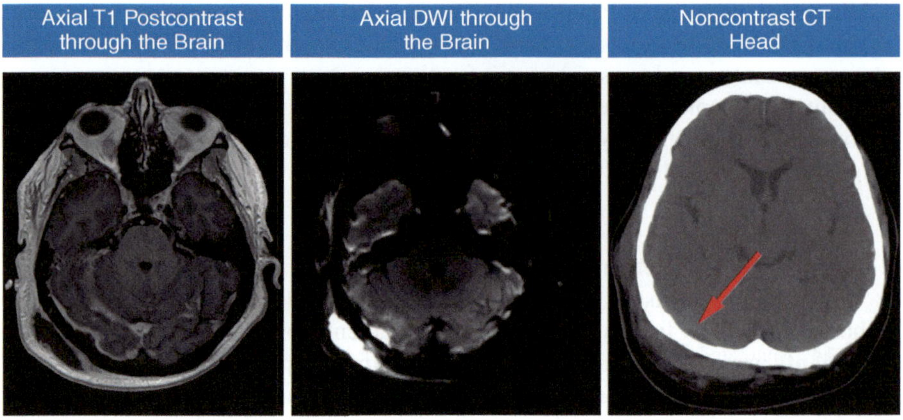

Fig. 2.3 A rather unusual case that began as minor trauma to the patient's right occipital scalp. The axial T1 postcontrast image demonstrates a right scalp organizing fluid collection and a right intracranial, extra-axial one too. There is thick pachymeningeal enhancement. The scalp and intracranial (extra-axial) fluid collections restrict diffusion. The CT reveals both fluid collections but also an intact calvarium. Surgical drainage revealed the collection to be purulent material (abscess) in the epidural space

so infrequent. In one instance, I have seen a venous epidural masquerade as the superior sagittal sinus (SSS), with the epidural fluid collection compressing the true sinus. Therefore, if a dural venous sinus is stripped away from the skull, it's likely a venous epidural.

What to look for:

- Lentiform hyperdense mass most often in the frontotemporal regions, especially after known trauma.
  - **Be aware that isodense epidural hemorrhages are possible if the patient has significant anemia.**
- Epidural hemorrhage in the posterior fossa is possible and extremely dangerous due to its limited capacitance.

What to document/report:

- Size of the epidural hemorrhage. As an additional pointer, when reporting values, pick **cm or mm and STICK WITH IT**. Don't mix measurements in a single report.
- Midline shift if present.
- Uncal encroachment (herniation) on the cerebral peduncles if present.
- Cerebellar tonsillar herniation if present.

3. Why it matters:

- Epidurals can grow and kill rapidly.
- Neurosurgical intervention is partly imaging-based.

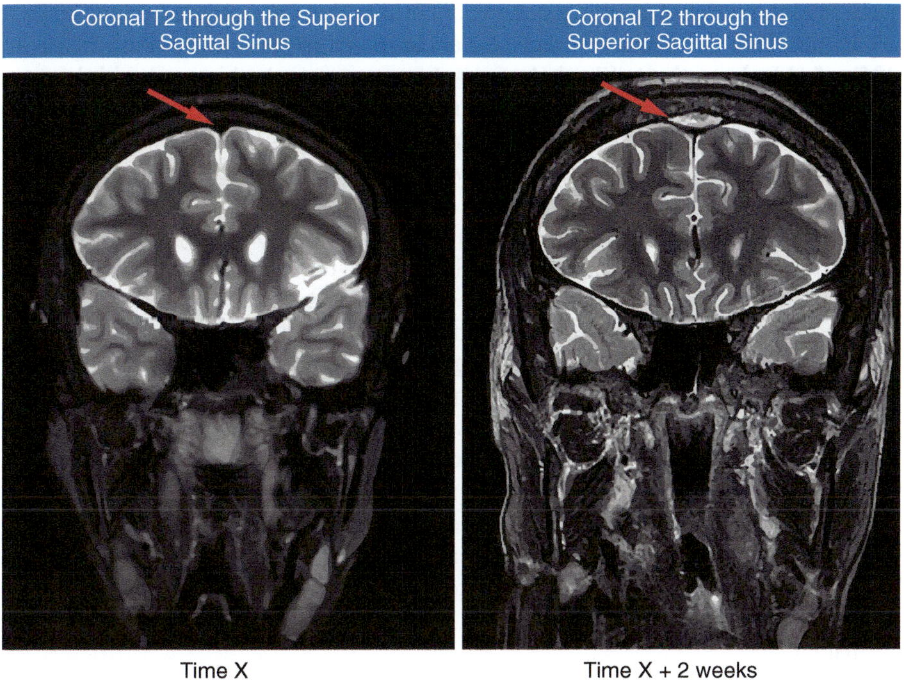

Fig. 2.4 In this figure, we see two coronal T2 images through the same plane separated by two weeks. The image on your left (Time X) is normal. The arrow denotes the SSS flow void. However, the image on your right two weeks later reveals a biconvex T2 hyperintense fluid collection *superficial* to the SSS (arrow). Diagnosis? Bone infarct in a patient with sickle cell anemia resulting in epidural hematoma. It is theorized that blood products are able to exude into the epidural space after infarct, in this case, effacing the SSS

- A Glasgow Coma Score (GCS) of less than 9 or focal impairment implies operative management.
- However, they will also intervene on epidurals greater than 30 mL in volume, affecting more than **5 mm midline shift**, or greater than 15 mm in thickness.
- Operative management may also depend on location and other variables, but volume greater than 30 mL appears to be the strongest surgical indicator [6] (Fig. 2.4).

2.4 Subdural Hemorrhage

2.4.1 Bottom Line Up Front

This subtype of hemorrhage will be the bread, butter, and bane of your existence as a neuroradiologist or neuro-enthused generalist. They are frequent, protean, and can present in the traumatic setting or spontaneously. They complicate around 30% of acute traumatic brain injuries and **can accompany mild head trauma in up to 20% of elderly patients** [7, 8]. As a result, you may be looking at multiple comparison exams when assessing a subdural hematoma. And to further complicate matters, rebleeding is definitely *a thing*, as the Millennials would say.

Five Diagnostic Pearls for the Resident/Fellow on Call:

- In a patient who is anticoagulated, suffers from a coagulopathy, and/or has significant degenerative cerebral atrophy, trivial trauma can cause a subdural hematoma.
- Subdural hematomas can be extraordinarily sneaky. In the acute setting, they are often bright, unless the patient has significant anemia. They will degrade by 1.5 Hounsfield units (HU) per day. That means they can be isodense to brain parenchyma in about 10 days [9]. In the **hyperacute setting**, they can be isodense to the overlying cortical gray matter. Look for sulcal effacement!
- Scrutinize every head CT through the vertex on axial imaging and back and forth in the coronal plane. Look for sulcal effacement and subtle asymmetries.
- Scrutinize the falx cerebri and tentorial leaflets on coronal imaging. Parafalcine subdural hematomas can be difficult to distinguish from normal blood.
- MRI can be helpful in assessing the acuity of a subdural hematoma. More on that later.

1. Acute Subdural Hemorrhage

   Hemorrhage into the subdural space can occur after traumatic injury of a bridging vein, spontaneously, through rupture of an arteriovenous malformation (AVM), or even after rupture of an intracranial aneurysm [10]. In the hyperacute setting, the blood will be around 30–40 HU. In a patient with normal hematocrit, the hemorrhage will consolidate and demonstrate 70–80 HU attenuation on a noncontrast head CT within hours. Unlike epidural hematomas, subdural hematomas can be holohemispheric **but do not cross the falx cerebri.**

   - The definition of acute subdural hematoma is **0–4 days**.
     - Hyperacute is occasionally described, which occurs **within minutes to hours of injury**. Hyperacute hemorrhage will be lower attenuating than acute because it has not had time to clot. I am harping on this topic because I have been burned before, and I don't want you to be (Fig. 2.5).
   - CT characteristics:
     - 70–80 HU attenuation
     - Irregular margins
   - MRI characteristics:
     - Hyperacute (intracellular oxyhemoglobin)
       - T1 isointense/T2 hyperintense
       - Heterogeneous appearance with swirling possible

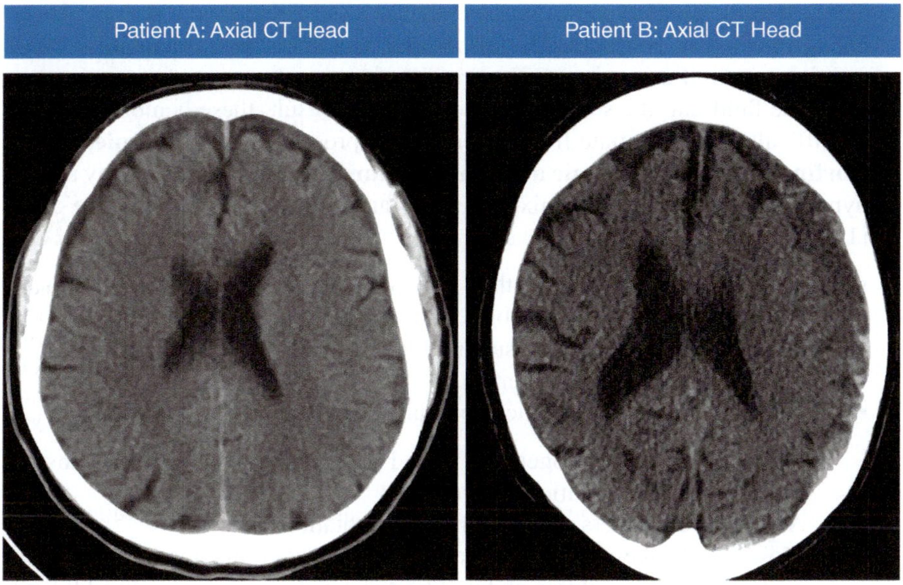

Fig. 2.5 Two axial CT images through two different patients. Patient A has subacute blood products in both frontal convexity subdural spaces. Patient B has subacute blood products in the right frontal convexity subdural space **and** acute-on-chronic left holohemispheric subdural blood products

  - Acute (intracellular deoxyhemoglobin)
    - T1 isointense/T2 hypointense

2. Subacute Subdural Hemorrhage

   Blood in the subdural space will degrade over time by about **1.5 HU per day**. Additionally, the chemical composition of the hemorrhagic contents will change, resulting in T1 shortening and variable T2 shortening and prolongation, respectively. **T1 shortening means T1 bright. T2 prolongation means T2 bright**.

- The definition of subacute subdural hematoma is **4–21** days.
  - This is further broken down into early subacute and late subacute.
- CT characteristics:
  - At around 10 days, the subdural can be isodense to cortical parenchyma making detection very difficult.
  - Degradation will vary and can be complicated by rebleeding.
- MRI characteristics:
  - Early subacute (intracellular methemoglobin)
    - T1 hyperintense/T2 hypointense
  - Late subacute (extracellular methemoglobin)
    - T1 hyperintense/T2 hyperintense

3. Chronic Subdural Hemorrhage

   Over time, the blood in the subdural space will become less conspicuous, but low-level rebleeding can occur. These can occur after lumbar puncture (or any cause of intracranial hypotension) and should be suspected if the patient experiences a persistent headache. To further complicate matters, granulation tissue can form as well as angiogenesis, creating functional neomembranes that can elaborate fluid into the subdural space [11]. As a result, these hemorrhages can be difficult to differentiate from other disease processes. **For example, Rosai-Dorfman disease can mimic a chronic subdural hematoma.** They may appear hypodense, isodense, or of mixed density, and they may calcify in rare cases [12, 13].

- Chronic is defined as greater than 21 days. The senior population is most at risk for developing this pathology.
- CT characteristics of chronic subdural hemorrhage:
  - Variable and potentially confusing.
- MRI characteristics of chronic subdural hemorrhage [14]:
  - T1 and T2 signal heterogeneity. Yeah, I know, there is no slam dunk answer here. You'll need additional sequences!
  - About three-quarters will have internal septations.
  - Most will display gradient/susceptibility-weighted imaging hypointensity
    - This varies in children and will be discussed in the abusive head trauma section.
  - Contrast enhancement will be thin and adjacent to the inner skull table in most cases. Occasionally, it can occur next to the brain surface.
  - Mass-like enhancement is possible.
  - Restricted diffusion in chronic subdural hematomas is possible, and empyema should be in your differential diagnosis! (Fig. 2.6)

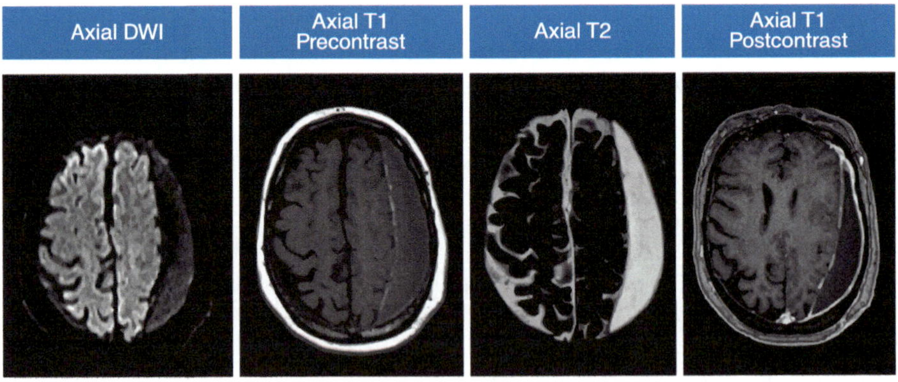

Fig. 2.6 Multisequence MRI images in a patient with a holohemispheric left subdural hematoma. Understand that establishing the age of subdural hematomas is difficult and can be confounded by re-bleeds, physiologic factors, and hygromas. The pertinent negatives you should report are: 1. Whether the collection restricts diffusion (here, it does not). 2. Mass effect, if any (here, there is no midline shift). And 3. Is there enhancement along the margins of the fluid collection (yes, there is). This helps rule out subdural empyema and establishes a chronic subdural hematoma

4. Subdural Hygroma

   I remember in my early training (crotchety old man voice) the confusion surrounding subdural hygromas. Are they subdural hematomas after a long wait? Are they CSF? The answer is yes. The best way to describe them is CSF in the subdural space due to anomalous—often posttraumatic—communication of the subdural space with the subarachnoid space. Again, to further augment the struggle sandwich that is neuroradiology, subdural hygromas can occur in the setting of bleeding, too, the creatively named *hematohygroma*. More recent data also suggest conversion of hygromas to chronic subdural hematomas [15].

- Subdural hygromas are associated with trauma, and they most often occur along the convexities. Posterior fossa hygromas are rare.
- CT characteristics of subdural hygroma
  - A “pure” hygroma will be isodense to CSF.
  - These can be confused with cerebral atrophy and expansion of the subarachnoid space. Here’s how you differentiate the two:
    - Do cortical veins traverse the CSF? If yes, it’s atrophy.
    - Are cortical veins displaced toward the brain surface? If yes, it’s subdural pathology.
  - What about a hematohygroma? The differential here is acute-on-chronic subdural hematoma and acute subdural hematoma [16].
    - Look for membranes. Membranes suggest an acute-on-chronic subdural hematoma. Otherwise, acute subdural hematoma should be on the differential diagnosis [17]!
- MRI characteristics of subdural hygroma:
  - T1 hypointense / T2 hyperintense essentially following CSF.

What to look for:

- On CT: Irregular hyperdense, extra-axial collection that crosses suture lines but does not cross the midline.
- On CT: Isodense material along the cerebral convexities can be difficult to detect. One must search for sulcal effacement, asymmetry, and subtle midline shift.
- On MRI: Assess axial T1 and FLAIR sequences for hyperintense asymmetry.
- On postcontrast imaging, look for septae/neomembranes that can help determine the chronic nature of a subdural collection.
- On postcontrast imaging, look for crossing vessels. If vessels run through a prominent subarachnoid space, you’re looking at cerebral atrophy with subarachnoid space expansion (Table 2.1).

Be aware that this chart is best applied to intraparenchymal blood. Subdural blood can deviate significantly from these guidelines. For residents, it may benefit you to learn the physics behind it for exam purposes. Hyperacute blood has oxyhemoglobin. Acute blood deoxyhemoglobin. Early subacute intracellular methemoglobin. Late subacute extracellular methemoglobin. Chronic

Table 2.1 A convenient reference chart for the MRI appearance of intracranial blood in various states

Time frame	T1 signal	T2 signal	CT density
Hyperacute: Within 6 hours	Intermediate	Bright	Hyperdense
Acute: 0–4 days	Intermediate	Dark	Hyperdense
Early subacute: 4–21 days	Bright	Dark	Isodense
Late subacute: 4–21 days	Bright	Bright	Iso-to-hypodense
Chronic: > 21 days	Dark	Dark	Hypodense near that of CSF

hemosiderin. Again, you must understand that blood will degrade at different rates in the intraparenchymal compartment vs the subdural space

What to document/report:

- Size of the subdural hemorrhage: measure the greatest thickness of the collection.
- Midline shift if present.
- Uncal encroachment (herniation) on the cerebral peduncles if present.
- Cerebellar tonsillar herniation if present.
- Skull table/surface enhancement or evidence of septal enhancement.
- Restricted diffusion if present.
- Peripheral enhancement if present.

Why it matters:

- Subdural hematomas can be confusing entities. Knowledge of their acuity will be important in clinical decision-making.
- Differentiating a subdural hematoma from a mimic, especially an empyema, of course, goes without saying in its importance.

Like epidural hematomas, the surgical management of subdural hematomas depends on clinical and imaging findings. If the subdural hematoma demonstrates a thickness greater than 10 mm or **a midline shift greater than 5 mm**, then surgical intervention is advised regardless of GCS. Alterations in the GCS or clinical change from baseline may precipitate surgery even if the hematoma does not meet the above sizes [18] (Fig. 2.7).

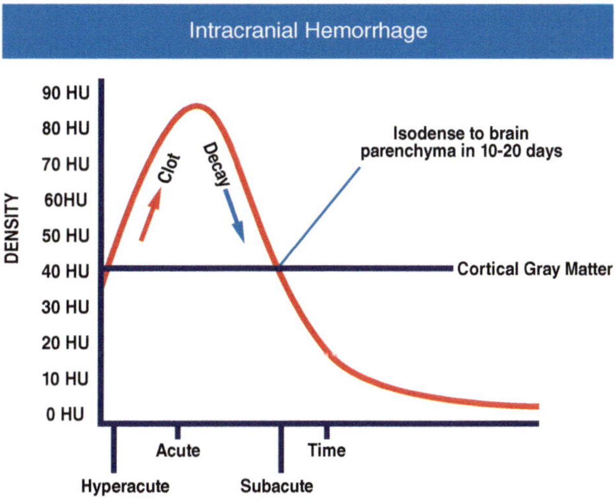

Fig. 2.7 This graph approximates the evolution of intracranial hemorrhage. The y-axis is density in Hounsfield units, and the x-axis is time. Note that blood doesn't immediately start as a floridly hyperdense collection. To convince yourself of this, look at the SSS and imagine blood flowing out of it into the subdural space in real time. It would be slightly hyperdense to cortical gray matter

2.5 Subarachnoid Hemorrhage

2.5.1 Bottom Line Up Front

Blood accumulation within the subarachnoid space, defined as the space between the arachnoid mater and the pia mater, is most often (85%) the result of aneurysm rupture in the nontraumatic setting. The textbook clinical history provided is that of "worst headache of my life," which will inevitably precipitate imaging. Given that blood is extremely irritating to the meninges, the patient may also present with neck pain and stiffness. However, subarachnoid hemorrhage is very different in the traumatic setting, where mild trauma can precipitate scant subarachnoid blood that often resolves with conservative measures. Unfortunately (and logically), patients with subarachnoid hemorrhage accompanied by concurrent additional types of hemorrhage have higher adjusted odds ratios for mortality [19].

Five Diagnostic Pearls for the Resident/Fellow on Call:

- Subarachnoid hemorrhage can be very faint on CT, especially along the convexities.
- A large volume of blood within the basilar cisterns is a rare presentation of traumatic subarachnoid hemorrhage, but a lethal one, as this finding is associated with vertebrobasilar artery rupture [20].
- Subarachnoid blood along the cerebral convexities is often a mild, self-limiting process in the setting of mild trauma.
- T2 FLAIR sequencing is very helpful for looking at subarachnoid blood, but one **should also include meningitis, patient O2 supplementation, and leptomeningeal carcinomatosis in the differential if sulcal FLAIR signal hyperintensities are seen.**

- You will experience cases of *chicken or egg* in which a patient presents with subarachnoid hemorrhage and evidence of neurotrauma. Did an aneurysm rupture, causing them to lose consciousness and fall? Or was it an aneurysm all along? The patient may need cerebral angiography or a lumbar puncture for definitive diagnosis.

1. Acute Subarachnoid Hemorrhage

You may be wondering: if there is an *acute* subarachnoid hemorrhage, can there be a chronic one? And the answer is no. Sort of. Due to the nature of blood's effect within the subarachnoid space, it cannot linger for long without killing the patient. Indeed, about 1 in 5 people who suffer from a subarachnoid hemorrhage will die before reaching the hospital. Further, if the patient re-bleeds, the mortality skyrockets to 78%. **This doesn't mean there aren't chronic stigmata of prior subarachnoid hemorrhages. Superficial siderosis, also known as hemosiderin staining, can be a useful clinical observation on gradient/SWI sequences**. Superficial siderosis may even explain hearing loss and other vestibulocochlear symptoms if it occurs in the posterior fossa with hemosiderin staining of the cisternal segment of the cranial nerve 7/8 complex. The pathophysiology of subarachnoid hemorrhage is quite interesting, multifaceted, and complex. See this source if you're interested in the nitty-gritty, but here are some highlights [21].

- When blood fills the subarachnoid space, the red blood cells lyse and release hemoglobin onto the brain parenchyma. The iron molecules then contribute to increased oxidative stress and free radical formation.
- Reduction of cerebral blood flow and loss of cerebral autoregulation occur, which precede vascular spasms that can themselves contribute to ischemia.
- There is a breakdown of the blood-brain barrier at the site of subarachnoid hemorrhage.
- Upregulation of inflammatory cytokines occurs as does the recruitment of immune cells.
- Blood products in the subarachnoid space result in the deposition of fibrin and fibrinogen.

Unfortunately for patients, the excruciating pain and ischemia are not the only problems that emerge after subarachnoid hemorrhage. Though the exact mechanism is unclear, hydrocephalus can develop in any phase of the process, further contributing to the morbidity of the disease, and about 30% will require permanent shunting for hydrocephalus after aneurysmal subarachnoid hemorrhage [22]. Interestingly, this author was unable to find much regarding the need for shunting after *posttraumatic* subarachnoid hemorrhage. However, since intraventricular hemorrhage is associated with an increased risk of shunting, one can surmise that the more severe the trauma, the greater the volume of subarachnoid blood (including intraventricular), and thus the greater the risk of shunting [23, 24].

What to look for:

- On CT: Make sure it is a noncontrast CT! Contrast staining can occur after the administration of IV or IA iodinated contrast, which will make assessing for hemorrhage much more difficult.
- If contrast staining occurs, repeat imaging in 24–48 hours is advised. Recent literature does note that **dual-energy CT can address this problem** using contrast subtraction [16, 25].
- Do your best to qualify the location of the hemorrhage: Is it the basal cistern or convexity?

What to document/report:

- Size of the subarachnoid hemorrhage.
- Is there an intraventricular component?
- Are the ventricles dilated? Have they changed from the prior study (if applicable).

Why it matters:

- Subarachnoid hemorrhage can be a relatively "benign" finding when seen in the setting of mild trauma.
- Larger volumes of blood, especially in the basilar cisterns, require urgent intervention due to vasospasm, cerebral ischemia, and hydrocephalus (Fig. 2.8).

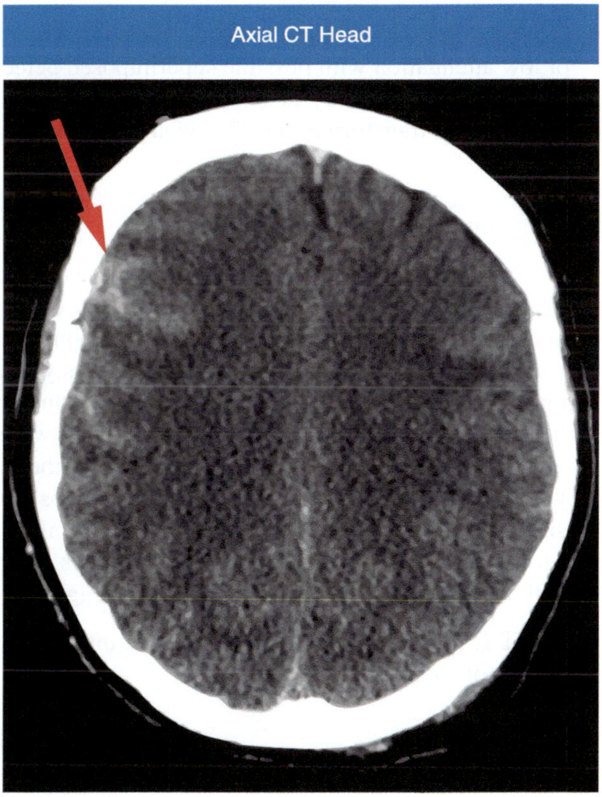

Fig. 2.8 Annotated axial noncontrast CT through the head demonstrates increased attenuation that follows the serpiginous/irregular configuration of the cortical sulci (arrow). This finding is concerning for subarachnoid hemorrhage. Recall that blood within the subarachnoid space flows along the cortical pial surface of the brain, which the brain finds *extremely* disagreeable

2.6 Intraparenchymal Hemorrhage

2.6.1 Bottom Line Up Front

Let's run with the human-brain-as-gelatin metaphor for the moment. The parenchyma is composed of gray and white matter, with the gray matter having a greater volume of vasculature and blood. The "gelatin" is seated in a hard skull (some harder than others), floating in a circulating volume of CSF. Blunt and piercing injuries to the gelatin will result in disruption of the substance of the brain, including blood vessels, whereby blood can pour into a traumatic cavity and expand.

Five Diagnostic Pearls for the Resident/Fellow on Call:

- Traumatic intraparenchymal hemorrhages occur just deep to the site of impact of the patient's skull. If you see a large scalp hematoma, look immediately deep to it for a skull fracture. Then go further. Scrutinize the "coup" brain parenchyma.
- Scrutinize the contrecoup brain parenchyma. Remember, a significant force transmitted through the skull will cause the opposite side of the brain to impact the skull.
- Carefully assess the most vulnerable regions of cortical parenchyma. These include the orbitofrontal gyri and temporal lobes due to their close proximity to the jagged and irregular surfaces of the frontal skull base, crista galli, and middle cranial fossae, respectively [26, 27].
- Other vulnerable regions to assess include the cerebellar and cerebral parenchyma as well as the midbrain.
- Cerebral contusions, also known as "brain bruises," will appear on CT as regions of low attenuation with possible superimposed petechial or frankly hemorrhagic material. They can vary in size. Do not mistake them for multiple infarcts with hemorrhagic transformation. That would be embarrassing.::nervous laugh::.

1. Acute Intraparenchymal Hemorrhage

   Know your history—clinical history that is, but don't hang your hat on provided clinical history alone! Spontaneous as well as traumatic intraparenchymal hemorrhage have overlapping presenting symptoms: nausea, vomiting, focal neurologic deficit, headache, etc. [28]. And as you can surmise, intraparenchymal blood in the setting of trauma is, well, due to trauma. Intraparenchymal hemorrhage in the *absence of trauma* invokes a wider differential diagnosis, and then, dear resident, there is the situation where there is no history. In these cases, your mettle will be tested because there are clues that can help make the diagnosis. Occasionally, we can have another chicken/egg scenario wherein a patient seizes, loses consciousness, and falls/crashes their car due to an underlying pathologic process like a brain metastasis (Fig. 2.9).

   - CT is the gold standard for assessing intraparenchymal hemorrhage regardless of cause.

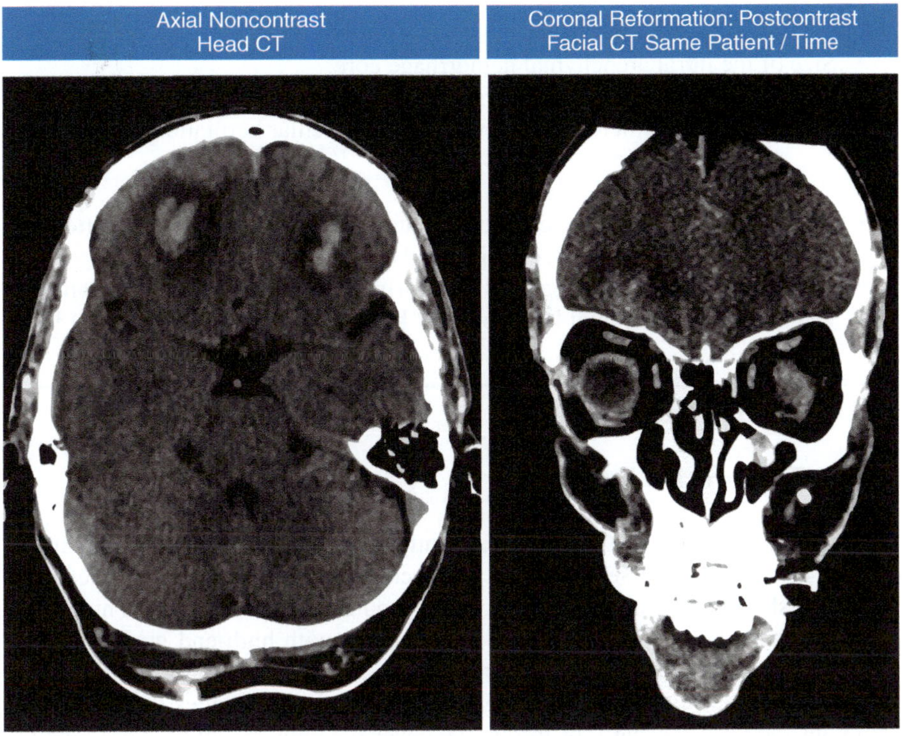

Fig. 2.9 On your left, you will see an axial noncontrast head CT through the basal frontal lobes. Bilateral hyperdense intraparenchymal hemorrhages are obvious. Take note of the surrounding low attenuation consistent with vasogenic edema. On your right is a coronal postcontrast image through the basal frontal lobes: same patient/same time. The hemorrhages are more difficult to see on the background contrast opacification. This effect can make small hemorrhages difficult or impossible to see

- Breakdown of the blood-brain barrier adjacent to the hemorrhage will invite worsening cerebral edema. In addition to the expanding hemorrhage, this edema can cause herniation, ventricular obstruction, and death.
- High blood pressure variability has been linked to hematoma progression and worse outcomes in one retrospective study [29].

What to look for:

- With a solid trauma history, start with a noncontrast head CT. Acute hemorrhage will leap off the screen at you.
- In cases where there is no history, start with a noncontrast head CT. Search for hemorrhage. If you find an intraparenchymal and/or subarachnoid hemorrhage, proceed to CT angiography of the head.
- Use MRI in cases where you suspect complexity. For example, if you see multiple circumscribed intraparenchymal hemorrhagic foci, keep hemorrhagic metastases or septic emboli in mind. Also, don't forget about AVMs.

What to document/report:

- Size of the intraparenchymal hemorrhage core.
- How much vasogenic edema surrounds this hemorrhage?
- Ventricular effacement, uncal herniation, or tonsillar herniation.

Why it matters:

- Intervention on intraparenchymal hematomas is based on volume, often calculated using the ABC/2 technique, and clinical exam.
- Multiple studies have suggested that early evacuation of both traumatic and spontaneous intracranial hematomas is beneficial to patient outcomes [30].

2.7 Diffuse Axonal Injury

2.7.1 Bottom Line Up Front

Diffuse axonal injury, or DAI as it's frequently referred, is a scary and tragic injury pattern. I know just one section prior we discussed the brain-as-gelatin metaphor. Now, let's shift over to a brain-as-personal-computer metaphor. I know a number of you nerds have built your own rigs before, replete with high-end graphics cards, rows of shiny DIMMS, liquid cooling systems, and cables galore. And if you haven't—that's ok. Just know that these systems are fragile. Knock the tower over, stub your toe on it too hard, or torque the cover off with too much force, and the screen stays blank when you press power. Indeed, substantial damage to one or multiple internal systems, with no external evidence of such, renders the computer nonfunctional. DAI is basically this, except for the human brain.

Five Diagnostic Pearls for the Resident/Fellow on Call:

- In a patient with DAI, CT of the brain can be completely normal, but the patient is comatose.
- If DAI is suspected, MRI has a much higher sensitivity in detecting the petechial hemorrhages associated with this injury. Gradient (T2*) sequences are useful, but susceptibility-weighted imaging (SWI) is better if available [31].
- Increase your suspicion for this injury if you are given the clinical history of a high-speed automobile accident, which is the most common cause [31].
- Pay very close attention to the number of lesions you see on MRI. **There is a dose-dependent correlation with lesion count and patient morbidity if they survive** [32]**.**
- Corpus callosum.

1. Acute DAI

   I said earlier that this injury pattern was tragic. Why? Because DAI is a severe type of traumatic brain injury that can leave the "lights off" permanently, even while the rest of the body heals. DAI is the result of stretching and occasionally shearing of axons that occurs with rapid acceleration, deceleration, and rotational forces on the brain. Since the gray and white matter densities are different, the axons within them are exposed to force differentials at the gray-white junctions. In short, DAI literally disconnects different regions of the brain by damaging association, projection, and commissural white matter tracts, which leads to Wallerian Degeneration [33] (Fig. 2.10).

- Pay specific attention to the gray-white junctions on both CT and MRI gradient/SWI sequences.
- Check the corpus callosum! **The body and splenium of the corpus callosum are very common sites of DAI** [34].
- **DAI of the brainstem portends much worse outcomes than the cortex** [35].
- Diffusion tensor imaging (DTI) research is a promising technique to better quantify and qualify DAI, with some evidence suggesting decreased fractional anisotropy in injured white matter tracts [36].

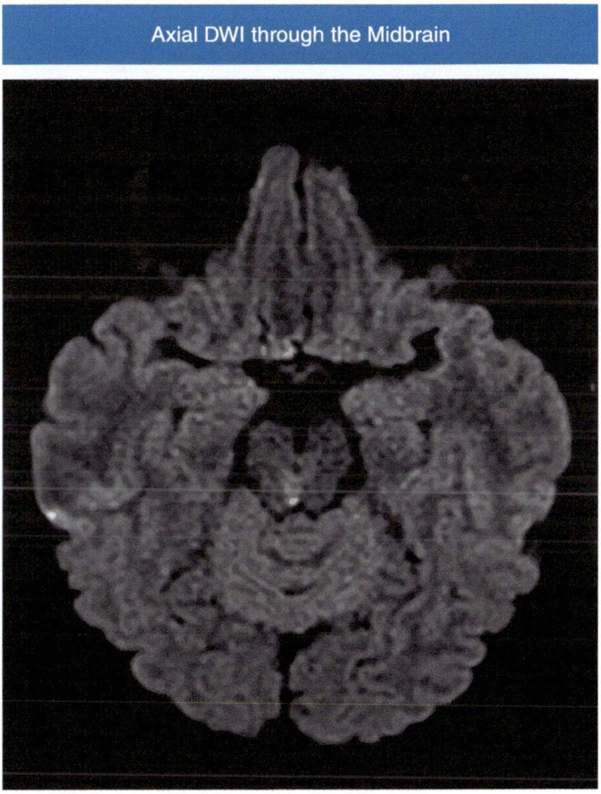

Fig. 2.10 Axial DWI through the midbrain shows a singular punctate focus of restricted diffusion within the dorsal midbrain. The patient had been struck by an automobile while on a scooter. The patient had multiple fractures and abrasions, but this was the only abnormality present within the brain. Clinical presentation? Internuclear ophthalmoplegia. This was an exceedingly rare case of traumatic axonal injury resulting in INO

What to look for:

- CT can be normal. Look for petechial hemorrhages (hyperdensities) at gray-white junctions.
- On MRI, search for SWI or gradient signal hypointensities. Carefully document where they are: cortical, corpus callosum, brainstem.
- Look for punctate FLAIR signal hyperintensities, which can be indicative of DAI without hemorrhage. Yes, non-hemorrhagic DAI is a thing.
- Look at your DWI sequence. DAI lesions, especially in the corpus callosum, can restrict diffusion, and they are correlated with worse outcomes [31].
- Go low. Scrutinize the medulla if it's visible. Injuries to the brainstem are much worse.

What to document/report:

- SWI/gradient hypointensities are indicative of microbleeds, which are a surrogate marker for axonal injury but not necessarily representative of the axonal injuries themselves [36]. Regardless, the greater the number of hypointensities, the worse the patient's prognosis.
- If you see SWI/gradient hypointensities in the brainstem, try to give an exact anatomic localization.

Why it matters:

- DAI is associated with a persistent vegetative state.
- SWI hypointensities in the mesencephalon, specifically the substantia nigra and tegmentum, portend worse long-term outcomes [37].
- The imaging findings can help generate a patient's prognosis and manage expectations (Fig. 2.11).

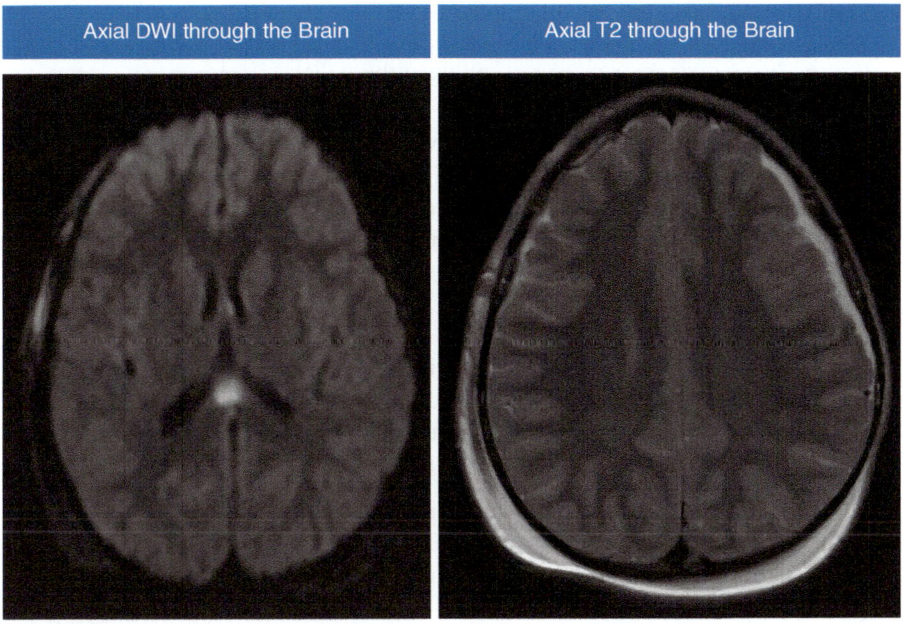

Fig. 2.11 A posttraumatic (fall) patient presents for imaging. The axial DWI reveals restricted diffusion within the mid-substance of the splenium of the corpus callosum. This finding is consistent with a cytotoxic lesion of the corpus callosum, also known as a CLOCC. The axial T2 image on your right reveals a left acute extra-axial hemorrhage and significant scalp swelling along the posterior calvarium. Commissural white matter fibers are distinctly vulnerable to axonal injury in the setting of trauma

2.8 Herniation and Obstruction Patterns

2.8.1 Bottom Line Up Front

Whether it be the setting of trauma, postoperative, spontaneous hemorrhage, or intracranial mass, one must be able to recognize the patterns of parenchymal herniation and ventricular obstruction. Both can be rapidly fatal.

Five Diagnostic Pearls for the Resident/Fellow on Call:

- Begin your assessment with symmetry in mind. In order to do this, the scan must be high quality: minimal patient motion artifact. If the patient is obliqued in the scanner, use the benefit of helical isotropic acquisition to create your own reformations.
- Comparison imaging is extraordinarily valuable in this setting. "Old gold" can prevent a lot of heartache, so do actually look at those comparison images.
- One of the more subtle findings is basilar cistern effacement. This finding can be challenging on CT because everything basilar is challenging on CT.

- Look for abrupt unilateral dilation of the temporal horns, which is consistent with entrapment. Similarly, look for abrupt ballooning of the frontal horns of the lateral ventricles, which can occur in cerebral aqueduct effacement or obstruction of the Foramen of Monro.
- **Don't wait to contact the ordering provider/care team if there is a significant change between examinations, even if the most recent examination doesn't meet the perfect definition of herniation. The patient could be moving rapidly in the wrong direction. It is your job to raise the alarm.**

1. Subfalcine Herniation

   The falx cerebri is a very unforgiving structure. It is tough, fibrous, and anchored firmly in place. Therefore, a medial shift of either hemisphere will force the squishy brain parenchyma beneath it. The first to go is the cingulate gyrus, which is why this is **also known as cingulate herniation**.

   - This is the most commonly encountered cerebral herniation.
   - Injury to white matter tracts, including the arcuate fasciculus, can occur due to direct mass effect. This has been reported to be a cause of various aphasias [38].
   - Branches of the anterior cerebral artery can be dragged beneath the falx and obstructed, leading to concomitant ACA-distribution ischemic infarct.

2. Uncal Herniation

   The uncus is a vulnerable patch of real estate in the anterior temporal lobe. With enough force, the uncus can be pushed below the tentorial notch and exert mass effect on the middle cerebral peduncle (midbrain). Of course, such an occurrence can lead to respiratory arrest (Fig. 2.12).

   - The following physical exam findings should raise your suspicion for uncal herniation:
     - Decreased consciousness or loss of consciousness
     - Blown pupil ipsilateral to herniation
       - CN 3 is immediately in the danger zone.
     - Contralateral hemiparesis
   - Increased intracranial pressure producing Cushing's triad:
     - Hypertension
     - Bradycardia
     - Apnea

3. Paradoxical Flap Herniation (Sinking Flap Syndrome)

   Following traumatic brain injury or other causes of intracranial hypertension, many patients will undergo decompressive craniectomy. This will allow the

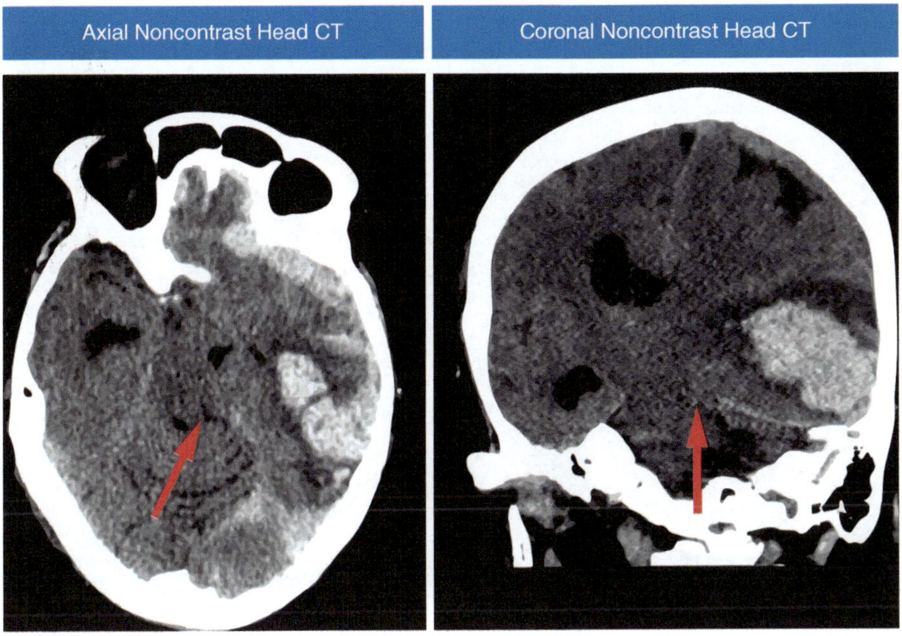

Fig. 2.12 Multiplanar noncontrast CT images display multiple significant findings in this one unfortunate patient. There is a large left temporal lobe intraparenchymal hemorrhage with surrounding edema. The arrows demonstrate uncal mass effect on the left middle cerebral peduncle (herniation). The coronal image (your right) shows severe left-to-right subfalcine herniation. There is unilateral dilation of the patient's right temporal horn consistent with entrapment

swelling brain some room to expand without generating a lethal mass effect on adjacent structures. In some patients, the relative position of the brain can change after craniectomy [39].

- The brain can herniate externally beyond the margins of the craniectomy.
- The brain can "sink" deeper, which is thought to be secondary to barometric changes.
- And the worst possible complication: the brain can herniate (paradoxically) away from the craniectomy defect, prompting an emergency situation (Fig. 2.13).

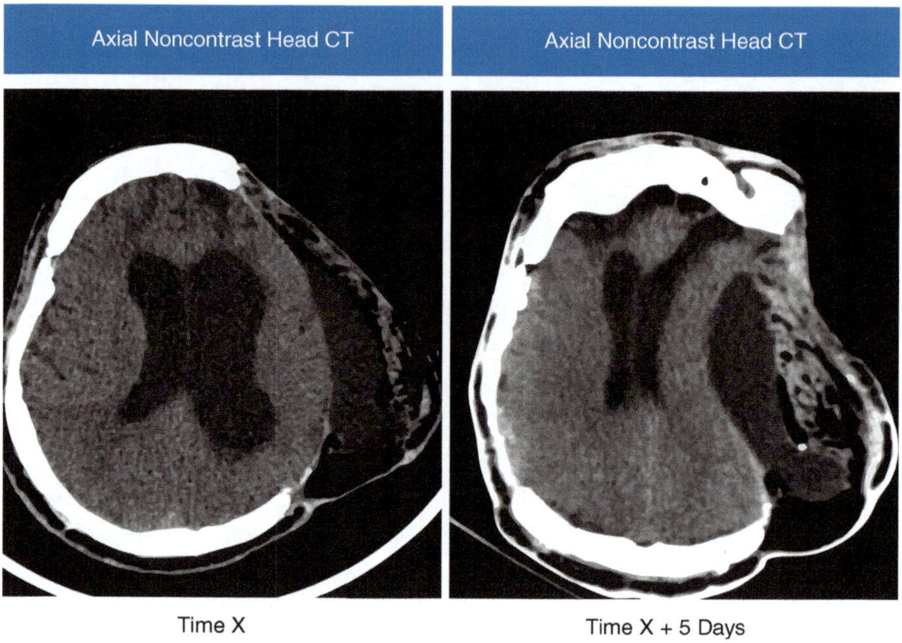

Fig. 2.13 On your left you see a postoperative axial noncontrast head CT. A large left craniectomy is seen with an autologous soft tissue graft overlying the defect. The image on your right was obtained after acute neurological changes were observed 5 days after the craniectomy. The image on your right shows sinking of the flap graft and paradoxical midline parenchymal herniation with ventricular effacement. This finding requires an immediate phone call

2.9 Craniofacial Trauma

Let's *face* it—heh—life as a modern-day Homo *sapiens* is pretty damn easy compared to life just a hundred years ago. And positively cush compared to life tens of thousands of years ago. In the preceding millennia, our ancestors and collateral bipedal buddies like Homo *habilis* had to navigate an incredibly violent landscape just to scrape a few bugs out of a log for dinner. Trauma was the norm in daily life—predatory cats, falls, predatory hyenas, hunting injuries, predatory bears, and other disagreeable hominids. You get the picture. So, it makes sense that we evolved some degree of protection in the form of the skull and paranasal sinuses to protect that gooey CPU known as the brain (Fig. 2.14).

- Skull Fractures
  - Scalp Hematomas
  - Depressed
  - Diastatic
    - Leptomeningeal Cysts
  - Basilar

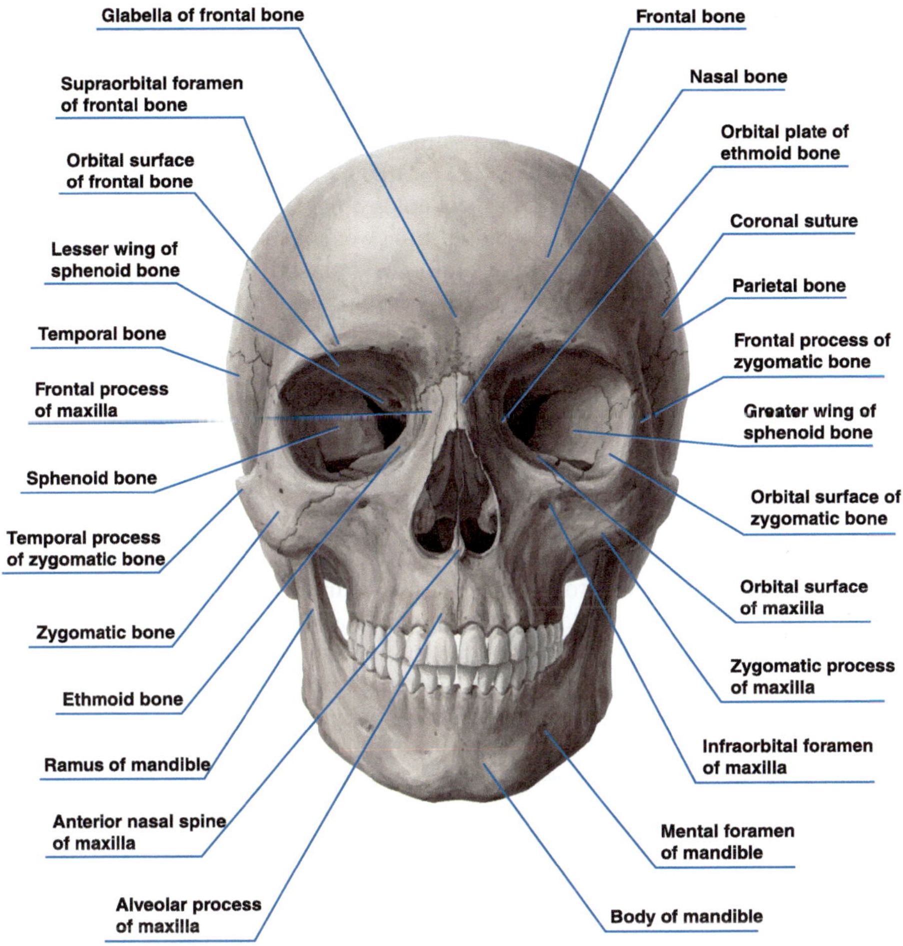

Fig. 2.14 An AP illustration of the human skull. Bro died with some perfect teeth, amirite? I recommend you memorize this anatomy as a foundation for further in-depth exploration later. Image courtesy of KenHub

- Facial Fracture Patterns
  - Buttresses
  - LeFort I
  - LeFort II
  - LeFort III
  - Zygomaticomaxillary (ZMC) Complex
  - Orbital
  - Naso-Orbito-Ethmoid (NOE)
  - Mandibular
- Orbital Soft Tissue Injuries
- Temporal Bone Fractures

  - Otic Capsule Sparing
  - Otic Capsule Disrupting
- Abusive Head Trauma (Formerly Nonaccidental Trauma)

Though life now does not compare to the savage violence experienced by our forbearers, there remains an ineffable risk of trauma in daily life. Furthermore, the trauma of today is exacerbated by our technology, namely, automobiles and industrial equipment, but fights with and without blunt weapons are a common occurrence.

As a radiologist on call, you must familiarize yourself with traumatic craniofacial injuries in order to give your patient the best chance of survival and to help mitigate morbidity.

2.10 Skull Fractures

2.10.1 Bottom Line Up Front

Trauma to the skull can cause fractures that propagate in random directions, which can be difficult to delineate from normal cranial sutures, especially in children. The location of a fracture can be an indication of the amount of force applied, as the frontal bone is very thick and the squamous temporal bone is very thin. Your first assessment of a skull fracture should be whether or not it is depressed.

Five Diagnostic Pearls for the Resident/Fellow on Call:

- **Do NOT settle for skull radiographs in the posttraumatic setting**. If the patient qualifies for a CT, either get the CT or have them transferred to a place with CT. Radiographs are obsolete and you **will** miss clinically significant pathology [40].
- The definition of a depressed skull fracture is inward displacement of a skull fragment greater than or equal to the width of the calvarium [41].
- Assuming you have the technological capabilities, ensure that spin and tumble 3D reformations of the skull and face are provided, especially for pediatric patients.
- Ensure that you have both soft tissue and bone kernel imaging available.
- When beginning your assessment, always scrutinize the scalp soft tissues, as scalp laceration in the setting of mild TBI can be associated with intracranial pathology in 20% of cases [42].

1. Scalp Hematomas

   There are three types of scalp hematoma: subgaleal, cephalohematoma, and caput succedaneum. While evaluating an adult trauma, I have found it to be not worthwhile to attempt to suss out which type of hematoma is present.

   - The subgaleal hematoma is found between the periosteum of the outer skull table and the galeal aponeurosis. It is **NOT** bound by suture lines and can expand significantly. **In newborns, this finding can be lethal, so absolutely do raise the possibility of it in a baby.**

- Cephalohematomas are found deep to the outer skull table periosteum and do not cross suture lines. This is a disease of infants. It most often resolves on its own. In the exceedingly rare cases where adults or older pediatrics are affected, retro-orbital extension is possible, which may cause vision loss and requires surgical intervention [43].
- Caput Succedaneum is a benign fluid collection external to the galeal aponeurosis that occurs due to birth trauma and is not pertinent to adult trauma.

2. Depressed Skull Fractures

It requires more energy to create a depressed skull fracture than a linear one. This additional energy creates comminution of the fracture fragments. Most cases of depressed skull fracture are managed operatively, but nonoperative treatment is sometimes preferred if depression is less than 10 mm, there is no dural injury, no hematoma, and no contamination of the wound, among additional criteria [44].

- Remember: It takes a lot of energy to depress the calvarium. Don't forget to assess the contrecoup brain.
- Be specific when describing fractures of the frontal sinuses. The injury will often necessitate surgery if the posterior skull table is involved, allowing intracranial communication and increasing the risk of CSF leak and meningitis [45].
- Whereas linear skull fractures are commonly seen in middle-aged individuals, depressed and comminuted skull fractures are more common in juveniles and the elderly [46].
- **Any skull fracture that crosses a dural venous sinus in an adult puts the patient at a 25–50% risk of a dural venous sinus thrombosis** [47, 48]. Make sure you mention this critical point in your impression.
- In the pediatric population, at least one study indicates venous sinus thromboses resolve spontaneously and do not require venous imaging unless there are additional clinical indications [49] (Fig. 2.15).

3. Diastatic Skull Fractures

There's not much to say about these injuries. They occur in children after injury to the cranial sutures, which have not yet fused. As my former noncommissioned officers used to say: *it would behoove you…* to learn the anatomy of the cranial sutures and the approximate ages of fusion.

- To further cement the point, 3D CT reconstructions of the skull are vastly superior to radiographs and even standard axial 2D CT images of the skull in finding diastatic skull fractures [50]. Demand them.
- Leptomeningeal cysts are also known as growing skull fractures, which are rare entities but most often occur in children less than 3 years of age. They are thought to be the result of cerebral parenchymal herniation into a fracture that slowly expands due to CSF pulsations. Patients who have fractures with greater than 4 mm of diastasis are at increased risk [51, 52] (Fig. 2.16).

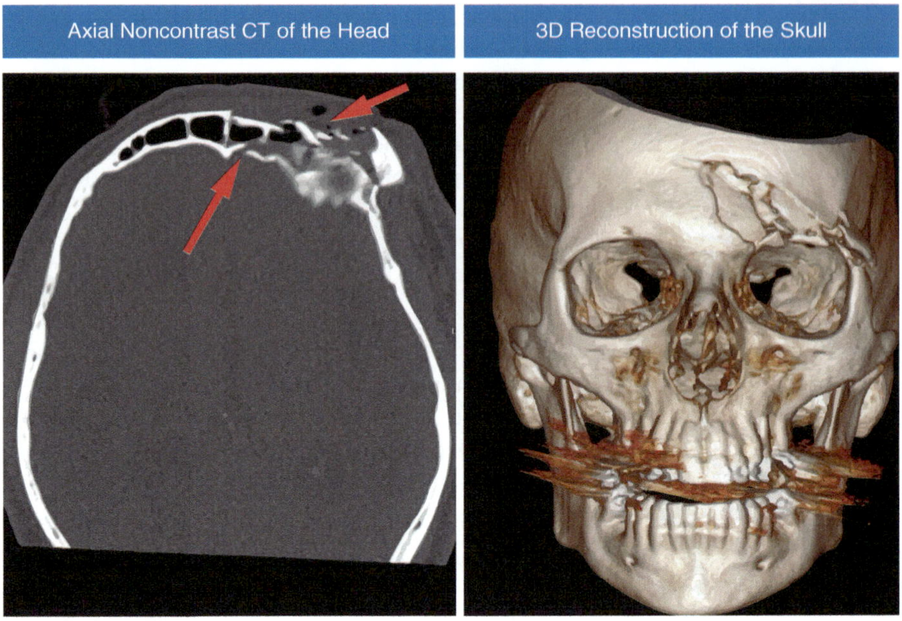

Fig. 2.15 The annotated axial CT image on your left reveals a comminuted fracture through the frontal sinus. Notice the inner and outer skull table is fractured. There is intracranial communication with the outside world, either directly (if the overlying scalp is deeply lacerated) or through inhaled air. The patient is at exceptional risk for CSF leak and intracranial infection. Notify the ordering physician of these findings

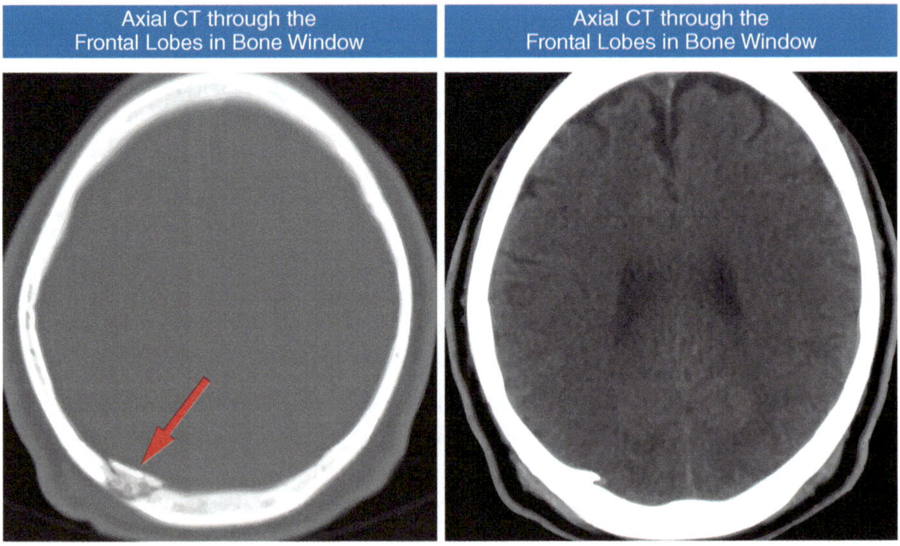

Fig. 2.16 An annotated axial CT image in bone window on your left reveals a few millimeters of depression of the inner skull table and irregularity of the diploe but preservation of the outer skull table. The image on your right is the same exam in the blood window, showing no hemorrhage

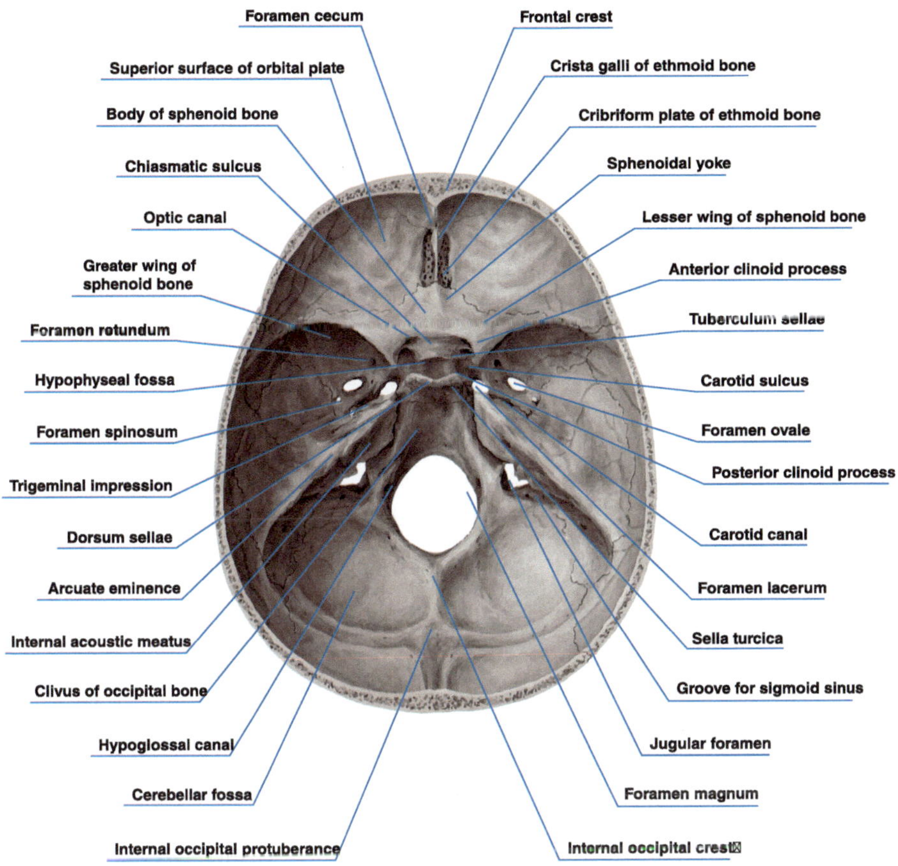

Fig. 2.17 An axial illustration of the skull base. Again, in order to be a competent radiologist, you must memorize every item in this image and learn what traverses them (if anything). Image courtesy of KenHub

4. Basilar Skull Fractures

   These are beastly injuries, inclusive of the dreaded temporal bone fracture, which will be discussed separately. Aside from the temporal bone, which comprises 40% of basilar skull fractures, the next most common sites are the frontal bone orbital roof, sphenoid, occiput, and ethmoid [53].

   - Common complications of basilar skull fractures include CSF leak, encephalocele, and meningitis [53].
   - Additional injury patterns are specific to the bone(s) involved and include facial nerve palsies, trigeminal nerve palsies, oculomotor nerve palsies, superior orbital fissure syndrome (SOFS) orbital apex syndrome (OAS), anosmia, and cavernous sinus thrombosis (Fig. 2.17, Table 2.2).

   A note on SOFS and OAS. These are two distinct ocular entities that can arise from craniofacial trauma. To complicate matters, cavernous sinus thrombosis can have significant overlapping symptoms. With respect to trauma, the orbital

Table 2.2 A list of the skull base foramina, their contents, and possible clinical manifestations of injury. Orbital apex syndrome and superior orbital fissure syndromes are further described below and very similar. The only major difference is orbital apex syndrome includes blindness

Skull bone	Foramen/canal/ fissure	Neurovascular structures	Possible complications
Frontal	Foramen caecum	Emissary vein	–
Ethmoid	Olfactory foramina of cribriform plate	CN I fibers	Anosmia, CSF leak
Ethmoid	Anterior ethmoid foramen	Anterior ethmoid artery, nerve, and vein	Significant bleeding, retrobulbar hematoma
Ethmoid	Posterior ethmoid foramen	Posterior ethmoid artery, nerve, and vein	Significant bleeding
Sphenoid	Optic canal	CN II, ophthalmic artery	Blindness Orbital apex syndrome (OAS)
Sphenoid	Superior orbital fissure	CN III, IV, V_1, VI	Superior orbital fissure syndrome (SOFS)
Sphenoid	Inferior orbital fissure	Infraorbital neurovascular, CNV_2 branches	Bleeding, trigeminal neuralgia
Sphenoid	Foramen ovale	CNV_3, meningeal artery	Bleeding, trigeminal neuralgia
Sphenoid	Foramen rotundum	CNV_2 maxillary	Trigeminal neuralgia
Sphenoid	Foramen spinosum	Middle meningeal artery	Bleeding
Sphenoid	Foramen lacerum (shared with temporal bone)	Nothing	Carotid artery injury associated with petrous-lacerum junction fractures
Sphenoid	Vidian canal	Vidian neurovascular	Parasympathetic: Xerostomia, dry eyes
Occipital	Foramen magnum	Brainstem, V4 vertebral arteries, CN XI, ant/post-spinal arteries, ligaments, meninges	Death, paralysis, vertebral dissection, vertebrobasilar thrombosis, CNXI palsy
Occipital	Jugular foramen (shared with temporal bone)	Jugular vein, CN IX, X, XI (pars vascularis gets X and XI; pars nervosa gets CNIX and inferior petrosal vein)	Venous thrombosis, CN palsies
Occipital	Hypoglossal canal	CN XII	CN palsy

apex and the superior orbital fissure are very close to one another, and there may be shared trauma to their indwelling structures and the Annulus of Zinn. CN II, III, IV, V1, and VI may be injured and present with palsies. But in short, the difference between OAS and SOFS is that OAS includes involvement of CN II, specifically vision loss, whereas SOFS does not [54].

What to look for:

- CT of the head will be the modality of choice in this setting. Ensure both soft tissue and bone kernels are available as well as 3D skull reformations.
- Begin your assessment with the soft tissues. Scalp hematomas, periorbital swelling, orbit and globe status, mastoid effusion status. These can give you initial clues of the injury pattern.
- Review the 3D skull images for gross deformity.

- Switch to axial bone kernel and search for fracture lines. If you detect any, annotate whether or not they cross a dural venous sinus. If they do, recommend CT venography. In a child, look for sutural diastasis.
- Review bone kernel in multiple planes. Pay specific attention to the basilar foramina above. Pay special care to the orbitofrontal and cribriform bones. Look for dehiscence that may suggest CSF leak or obvious encephalocele.

What to document/report:

- Size and location of the scalp hematoma. Does it cross sutures?
- Location of skull fracture, diastasis (if pediatric), depression, does it cross a dural venous sinus?
- Are any basilar skull fractures present? Which skull base foramina or canals do they involve?

Why it matters:

- CT venography is critical for picking up dural venous sinus thromboses.
- Fracture lines that approximate, cross, or deform any foramen/canal can hint at neurovascular injury. Describing this can help mitigate palsies, dissections, thromboses, pain, and weakness.
- Injuries of the orbital apex can cost a patient their vision. Intervention with steroids and surgical decompression can prevent this outcome. More on this later.
- In young pediatric patients, a subgaleal hematoma can be fatal.

2.11 Facial Fracture Patterns

2.11.1 Bottom Line Up Front

The word *backpfeifengesicht* is German, meaning one who possesses a face that you just want to punch. Not that I encourage gratuitous and random violence, but there exists in our world a lot of trauma aimed at faces. Sometimes fists. Sometimes baseball bats. Sometimes dashboards or airbags. As a radiologist, you will see facial fractures very often. These injuries can hint at deeper injuries in the head, or they may occur in isolation. Regardless, an organized, clean description is essential in proper patient disposition, because, although they may not be fatal, facial fractures can be very morbid.

Five Diagnostic Pearls for the Resident/Fellow on Call:

- Head and facial bone CTs will occur together. Given that helical CT data acquisition is isotropic, these studies should be reformatted separately in bone and soft tissue kernels and in multiple planes.
- Crafting a useful report in this setting is an art as much as it is a science. Robotically listing every fracture is a common occurrence in radiology, which leaves the surgeons no choice but to sift through extraneous content and look at

the imaging themselves. Your job is to add value, not confuse, annoy, or sink into irrelevance.

- Struts and buttresses are concepts in the developing nomenclature that have surgical implications. You must learn them.
- Nasal bone fractures are the most common in all settings; however, mandibular, maxillary, and orbital fractures are the top three (in that order) requiring surgery [55]. This does not mean that nasal bone fractures do not require surgery. If a nasal bone fracture is accompanied by a septal hematoma, obstructs the airway, or results in cosmetic deformity, then it requires surgery.
- There are plenty of important foramina in the face too. Learn them.

1. Buttresses

   This terminology can sound strange when applied to a human being, and so too can *struts*. However, it is of sound logic, as it describes the support structures of the face and how they articulate with the cranium. Moreover, these buttresses are actually thicker bone than the surrounding structures, lending themselves to support [56]. With respect to *thirds*, the face can be separated into upper, middle, and lower thirds. The superior orbital rims define the superior border of the middle face. The maxillary teeth define the inferior border of the middle face. Now it's easy to picture the rest.

   - There are four vertical buttresses that connect the face to the skull base. From lateral to medial:
     - Posterior vertical mandibular (posterior mandible)
     - Lateral maxillary (lateral orbit and maxillary sinus walls)
     - Posterior maxillary (pterygoid plates)
     - Medial maxillary (medial orbital and maxillary sinus walls)
   - There are four horizontal buttresses. From superior to inferior:
     - Upper transverse maxillary (orbital floor)
     - Lower transverse maxillary (includes the hard palate)
     - Upper transverse mandibular (mandibular alveolar process)
     - Lower transverse mandibular (inferior mandible)
   - There is another classification system for buttresses that includes three horizontal and three vertical. This system excludes the mandibular segments [57].
     - Vertical: nasomaxillary (medial maxillary)
     - Vertical: zygomaticomaxillary (lateral maxillary)
     - Vertical: pterygomaxillary (posterior maxillary)
     - Horizontal: superior orbital rim (frontal bar)
     - Horizontal: inferior orbital rim (upper transverse maxillary)
     - Horizontal: maxillary alveolar rim/hard palate (lower transverse maxillary)
   - When we discuss facial fractures, we will describe the buttresses involved, their displacement, and other key anatomic landmarks such as foramina (Fig. 2.18).

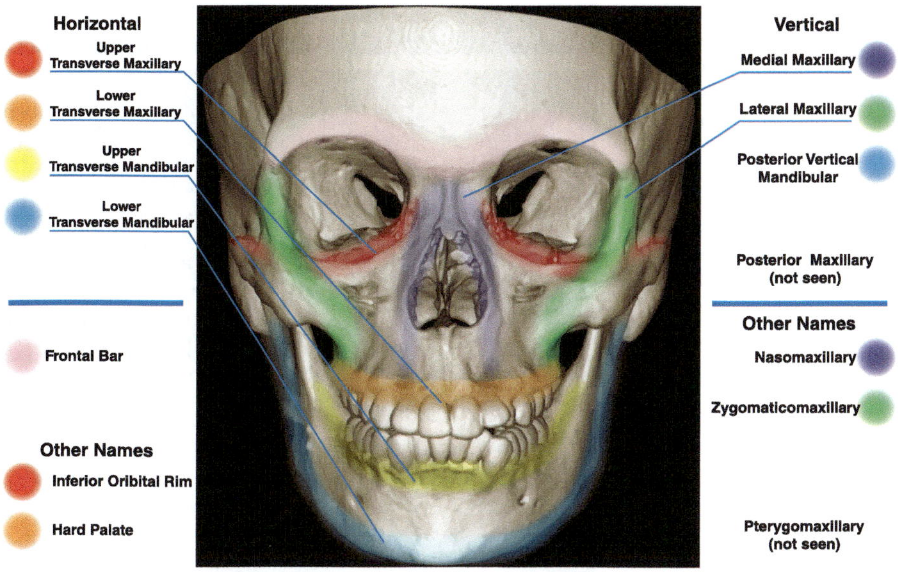

Fig. 2.18 A coronal 3D skull overlaid with colors corresponding to the buttress system. Other names for the same structures are noted at the inferior half of the image. The pterygomaxillary (posterior maxillary) buttress is not seen. For those with color perception difficulties, the most superior buttress across the superior orbit is the **frontal bar**. The **upper transverse maxillary** corresponds to the inferior orbital rim. The **lower transverse maxillary** corresponds to the hard palate. The **medial maxillary** corresponds to the nasomaxillary. The **lateral maxillary** corresponds to the zygomaticomaxillary

2. LeFort I Fractures

   The LeFort Classification System for facial fractures was developed ages ago by a Frenchman who smashed cadaver faces and documented fracture patterns. This classification system is still used by surgeons. The namesake is also carried over to LeFort osteotomy surgeries for cosmetic facial corrections, which is beyond the scope of this text. ***All LeFort injury patterns involve fractures of the pterygoid plates (posterior maxillary / pterygomaxillary buttress). However, not all pterygoid plate fractures reflect a LeFort injury.***

   - Involved buttress.
   - Posterior maxillary:
     - Inferior segment of the medial maxillary
     - Inferior segment of the lateral maxillary
   - Extension through pterygoid plates (like all LeFort fractures).
   - Effectively separates the hard palate from the face, creating a “floating palate.”
   - Nasal septal involvement occurs in bilateral injuries.
   - Anterior malocclusion and maxillary mobility can be noted on physical exam (Fig. 2.19).

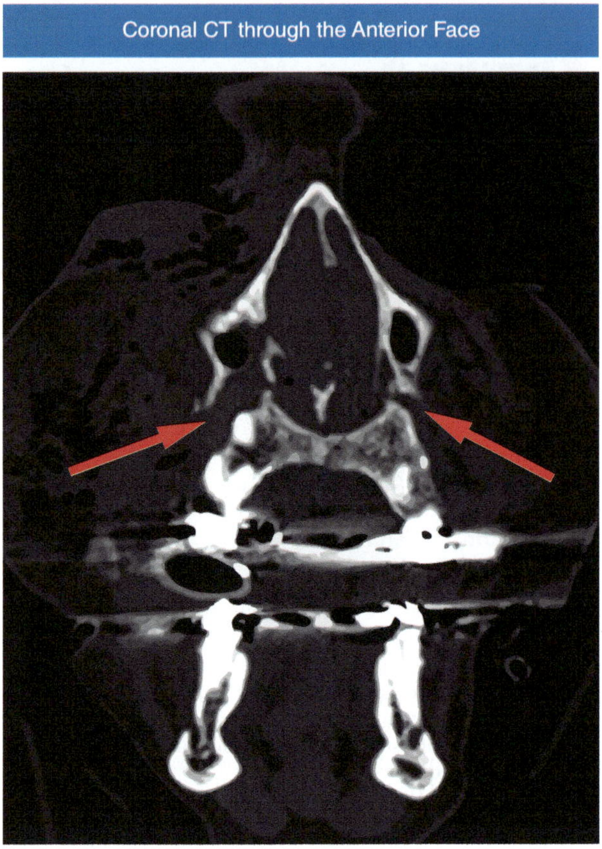

Fig. 2.19 An annotated coronal CT image through the anterior face at the piriform aperture. The two arrows point to bilateral lateral nasal aperture fractures. This injury pattern should raise suspicion for a floating palate (LeFort I injury)

3. LeFort II Fractures

    This fracture is also known as a pyramid fracture by the shape of the resulting maxillary facial fragment, with the maxillary teeth acting as the base of the pyramid. The fracture will involve the medial orbit, orbital floor, and zygomaticomaxillary suture.

- Multiple buttresses are involved in this fracture, including:
  - Posterior maxillary
  - Medial maxillary
  - Lateral maxillary (perhaps a small lateral portion of this buttress)
  - Upper transverse maxillary
- Often caused by trauma to the middle or inferior maxilla.
- If you find one LeFort fracture, keep looking for others. There can be multiple on a single patient, and they need not be symmetric (Fig. 2.20).

"Pyramidal" Configuration of a LeFort II Fracture

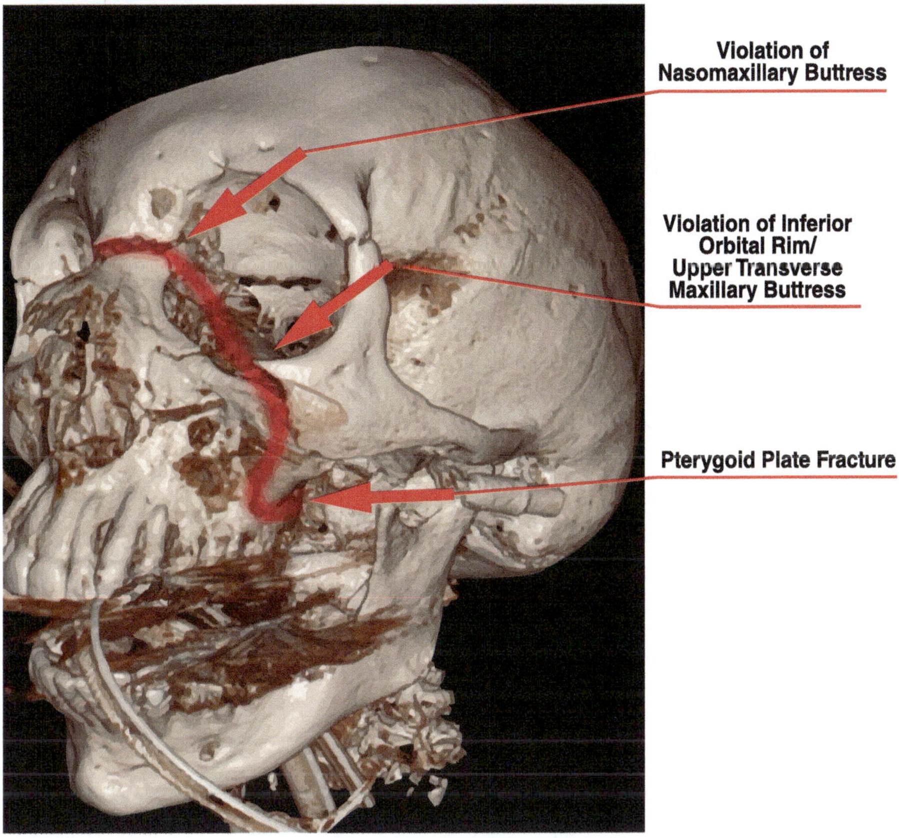

Fig. 2.20 This obliqued and annotated 3D reconstruction demonstrates nicely a unilateral component of this unfortunate patient's left LeFort II fracture. Indeed, this patient also had a left LeFort I fracture. Can you see it?

4. LeFort III Fractures

   Craniofacial dissociation is the apt descriptor of this fracture. They are associated with extreme blunt force impacts to the nose and upper maxilla. Since LeFort IIIs involve the skull base, there is also a risk of CSF leak.

   - Multiple buttresses are involved in this fracture pattern, too, including:
     - Posterior maxillary
     - Medial maxillary
     - Lateral maxillary
     - Upper transverse maxillary
   - This fracture is the only one that involves the zygoma proper (Fig. 2.21).

   Takeaway Points for LeFort Injuries:

   - Not all pterygoid plate fractures indicate a LeFort injury, but all LeFort injuries require pterygoid plate fractures (Fig. 2.22).

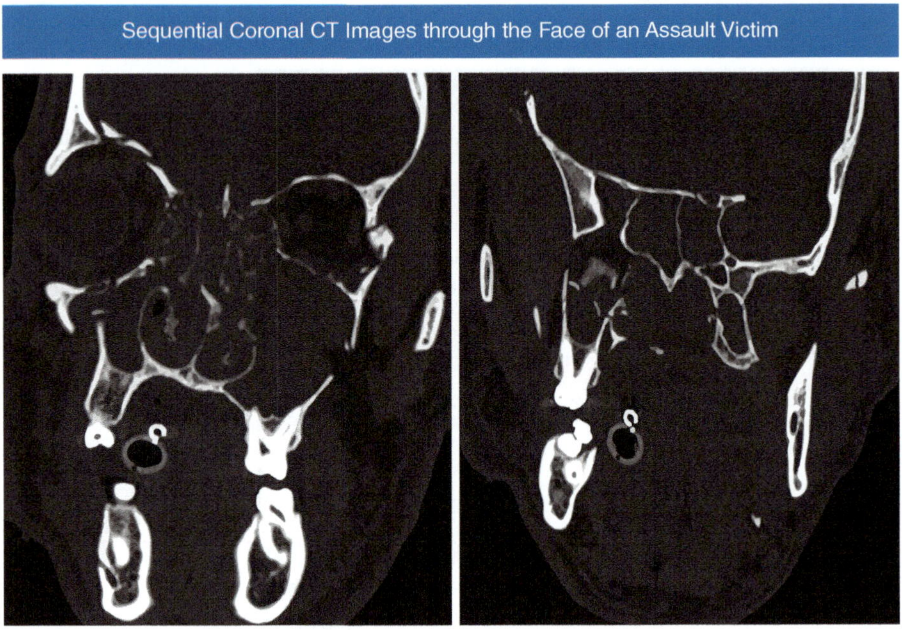

Fig. 2.21 Coronal CT images through the face at the level of the maxillary sinuses (left) and pterygoid plates (right). This poor soul sustained terrible facial trauma with multiple fractures. LeFort 1, 2, and 3 injuries are present bilaterally in this patient

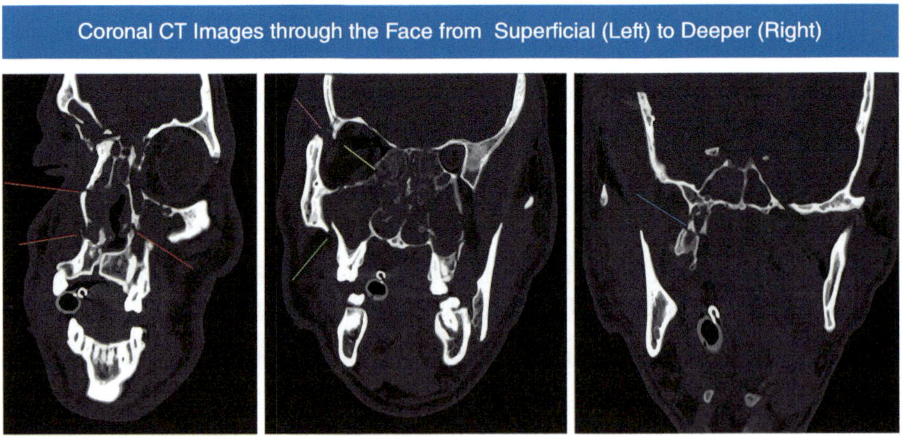

Fig. 2.22 These coronal sequences are arranged most anterior on your left and most posterior on your right. The red lines demonstrate fractures through the margins of the nasal aperture. The middle image demonstrates fractures through the maxillary sinus, medial inferior orbit, and zygoma. Finally, the right image shows a fracture through the pterygoid plate. All three LeFort patterns are seen here

- **Two clever authors helped streamline recognizing these injuries by describing unique components of each** [58]:
  - **LeFort I: anterolateral margins of the nasal fossa**
  - **LeFort II: inferior orbital rim**
  - **LeFort III: zygomatic arch**

5. Zygomaticomaxillary (ZMC) Complex Fractures: Sometimes "tripod" or "tetrapod."

   This is the second most common facial fracture after the lowly nasal bone fracture. It is caused by impacts on the malar eminence—think left or right hooks or rotated head onto a dashboard at high speed. Zygoma comes from the Greek word for "yoke," which is a wooden crosspiece that bridges a pair of animals. This is fitting because the zygoma, which also forms the lateral orbital wall, links the face and calvarium via four sutures

   - Involved buttresses:
     - Upper transverse maxillary (inferior orbital rim)
     - Lateral maxillary (zygomaticomaxillary)
   - Four sutures "yoke" (bridge) the face and calvarium
     - Zygomaticomaxillary
     - Zygomaticofrontal
     - Zygomaticosphenoid
     - Zygomaticotemporal
   - ZMC fractures can lead to cosmetic deformity via asymmetry and enopthalmos or exopthalmos.
   - Involvement of the posterolateral wall of the orbit is common and can develop into SOFS, necessitating surgical decompression [59, 60].
   - Fracture of the arch can impinge upon the temporalis muscle or masseter, causing trismus. ZMC fractures can also commonly injure the infraorbital nerve [61].

6. Orbital Fractures

   The globes sit comfortably in their sockets—their orbits—padded by fat and secured in place with a complex network of connective tissues. The orbits are cone-shaped and comprised of seven distinct bones. Fortunately, we as radiologists don't have to describe each bone in an orbital injury. Rather, discussion is centered on the four walls and the orbital soft tissue contents (Fig. 2.23).

   - Four walls of the orbit and associated buttresses:
     - Medial orbital wall (medial maxillary/nasomaxillary buttress)
       - Find the lamina papyracea on axial CT.
     - Inferior orbital floor (inferior orbital rim/upper transverse maxillary)
     - Lateral orbital wall (lateral maxillary/zygomaticomaxillary)
     - Orbital roof (superior orbital rim/frontal bar)
   - Orbital floor fracture:
     - Orbital floor blowout fracture is the most common orbital fracture.
     - Assess the size of the defect.

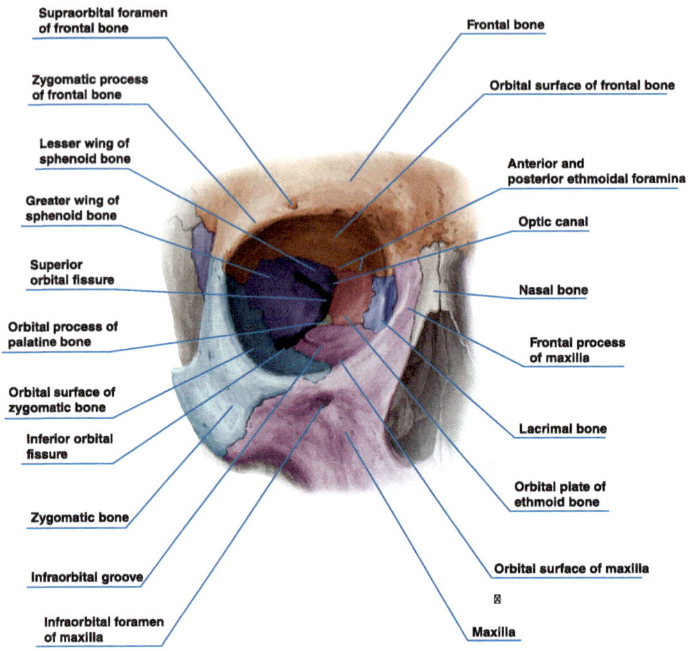

Fig. 2.23 A magnified AP illustration of the orbit with the various bones that comprise it. You know the drill. ***Learn to swim… through the relevant anatomy.*** Image courtesy of KenHub

  - Assess the amount of fat displaced into the maxillary sinus.
  - Assess whether the inferior rectus is speared by a bone fragment or herniating into the maxillary sinus (Fig. 2.24).
- Medial orbital wall (lamina papyracea) fracture
  - Similar mechanism as the orbital floor fracture, except the contents herniate inward into the sphenoid sinus. The medial rectus is at risk here.
- Lateral orbital wall fracture
  - Rare as this is the strongest component of the orbit.
- Orbital roof fractures
  - High-energy injury that is most often accompanied by other facial fractures [62].
  - May communicate with the anterior cranial fossa.
  - Risk of dural injury and CSF leak.
  - Meningitis is also a risk (Fig. 2.25).
- Orbital apex
  - The anatomy of the orbital apex is complex, but your job is simple in the traumatic setting. Is there a bone or hematoma deforming/effacing the orbital apex?
  - The SOF is a gap between the greater and lesser wings of the sphenoid.

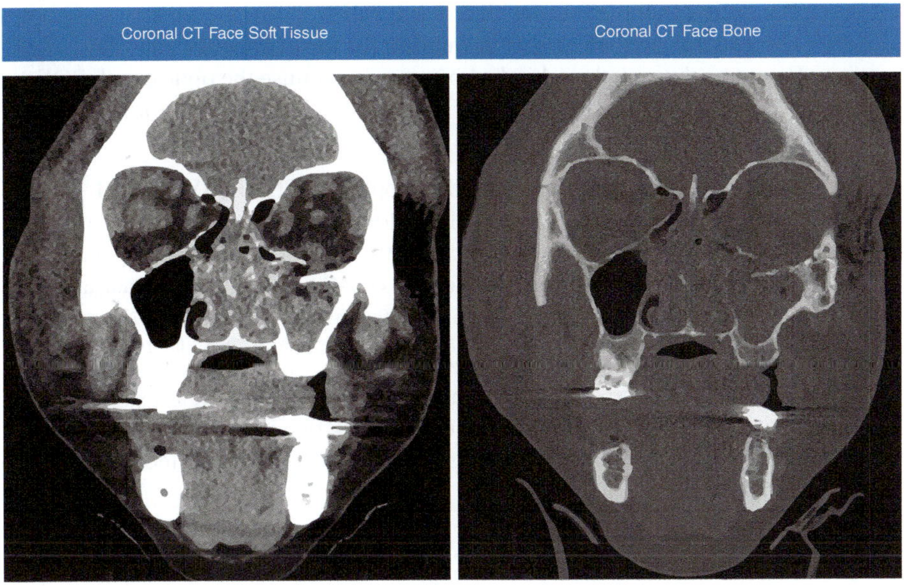

Fig. 2.24 Coronal images of the orbits in soft tissue and bone window reveal a depressed left orbital fracture and fracture of the left lamina papyracea. There is medial orbital fat content herniation. Also, look at the left inferior rectus muscle. It appears rather uncomfortable with a bone fragment at its medial margin

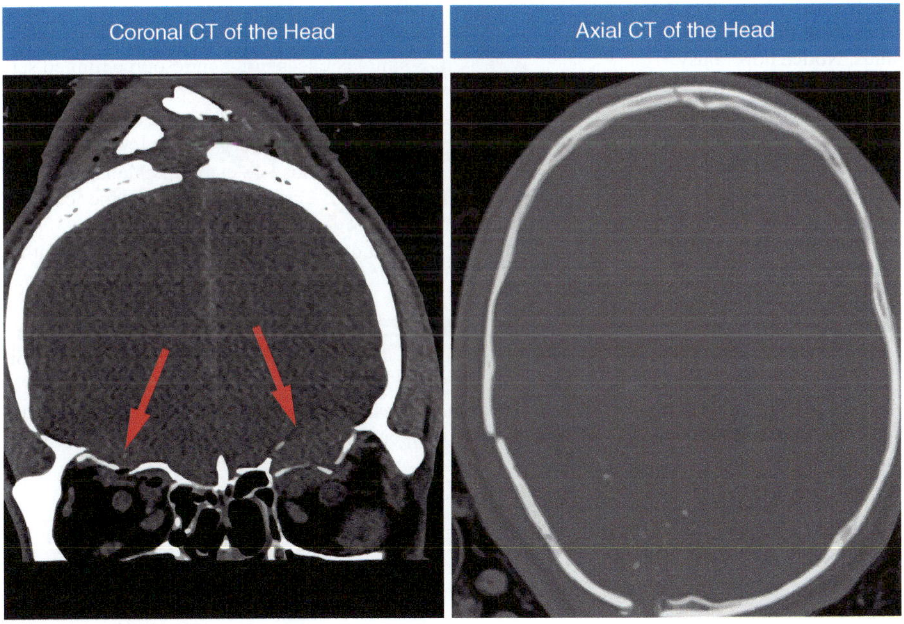

Fig. 2.25 The annotated image on your left shows both superior orbital rims collapsed into the orbital compartments, so-called orbital "blow-in" fractures. You'll also notice bone fragments displaced from the calvarium with significant soft tissue swelling and hemorrhage. The image of the calvarium on your right shows skull fractures and a posterior bullet wound. This patient was shot at close range in the back of the head

- The optic canal is superomedial to the SOF.
- The annulus of Zinn acts as the anchor for the extraocular muscles.
- Fractures through the orbital apex can compromise the optic canal, SOF, or both, leading to orbital apex syndrome (Jacod syndrome) or SOFS (Rochon-Duvigneaud syndrome).
  - Orbital apex syndrome: ophthalmoplegia, CN III, IV, and VI palsies, ptosis, forehead numbness, vision loss.
  - SOFS: Very similar to orbital apex syndrome, except the patient is spared vision loss. Complicates 0.3–0.8% of facial traumas [63] (Figs. 2.26 and 2.27).

7. Naso-Orbito-Ethmoid Fractures [64–66]

   The NOE region of the face is a confluence of multiple structures. This injury pattern, like others, requires a high-energy smack dab in the middle of the face. Surgery is often required in this setting due to the significant cosmetic deformity that can result.

   - Involved buttresses
     - **Medial maxillary (nasomaxillary)**
     - Frequently associated with frontal sinus fractures (frontal bar)

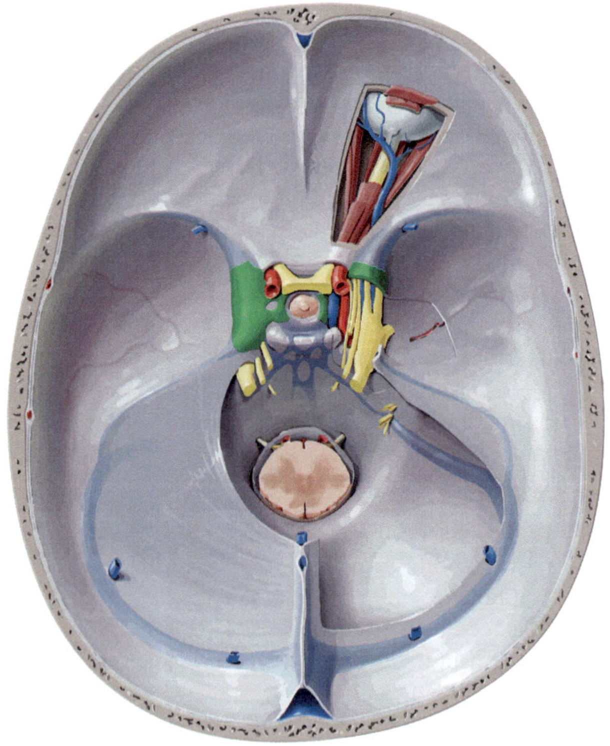

Fig. 2.26 An axial illustration through the skull base. The green is the territory of the cavernous sinus. Notice how “busy” the regions of the optic canal and superior orbital fissure are. **The following structures are found within the cavernous sinus: CN 3, 4, V_1, V_2, 6, and the internal carotid artery. CN 6 is most medial. Memorize this.** Image courtesy of KenHub

Fig. 2.27 A magnified AP illustration of the (common tendinous) Annulus of Zinn, the superior orbital fissure, optic canal, and inferior orbital fissure. "Big yellow" in this image is the optic nerve. Image courtesy of KenHub

- The critical soft tissue structure here is the medial canthal tendon. Ironically, it cannot be seen on CT, but surgeons *will* ask you if there is involvement.
  - On physical exam, NOE fractures can result in:
    - Epiphora (watery eyes, tearing)
    - Flattened Nose
    - Telecanthus (increased intercanthal distance)
    - CSF rhinorrhea
  - Look for a fracture that involves the **frontal process of the maxilla**. This is something the surgeons definitely want to know as it suggests medial canthal tendon injury.
- Correlative point: LeFort III fractures must also include a NOE component (Fig. 2.28).

8. Mandibular Fractures [67, 68]

   As the old saying goes: *talk shit, get hit.* And depending on the severity of the hit, the mandible can fracture in a number of places. Of course, there are other ways to suffer mandibular trauma aside from interpersonal violence, namely, falls and automobile accidents. Regardless of the mechanism, one should remember that the mandible is a ring osseous structure, like the pelvis, so if there is a fracture in one place, there will likely be a fracture in another.

   - Involved buttresses:
     - Upper transverse mandibular
     - Posterior vertical mandibular
   - There are a bunch of classification systems for mandibular fractures. I don't recall using any of them in my dictations. Rather, you may find it useful to review them to understand what the surgeons consider important.

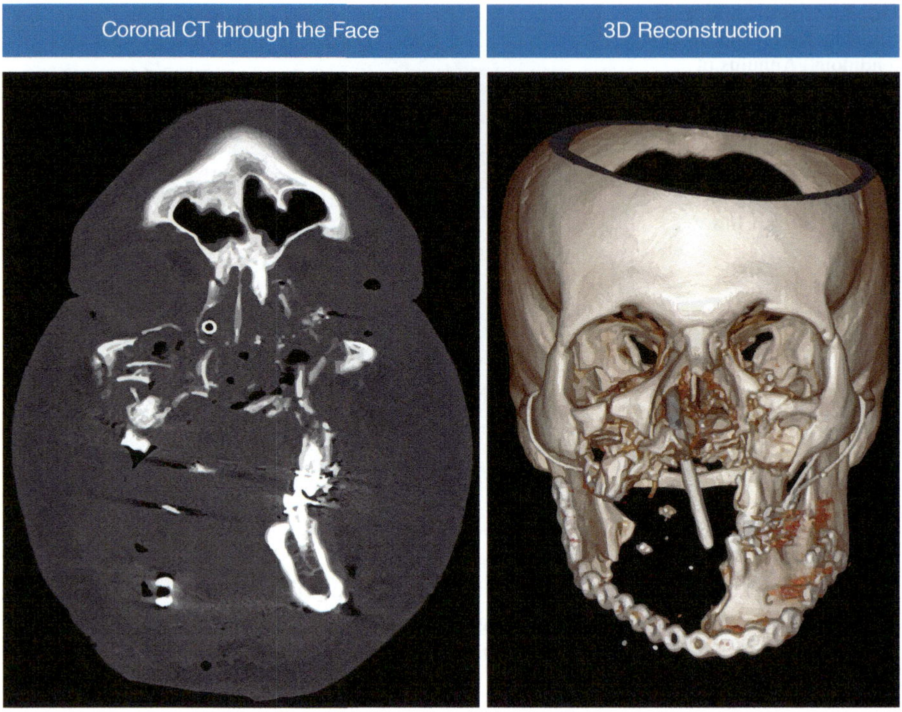

Fig. 2.28 Coronal CT on your left and 3D reconstruction on your right of a patient shot in the face at close range. There are many fractures here. The naso-orbito-ethmoidal (NOE) complex injury pattern involves both medial orbital walls, the ethmoid bone, and the nasal bones. The left side is worse. In your reports, make special mention of the degree of comminution of the medial orbital walls. Why? Because it gives the surgeon an idea of whether the **medial canthal tendon** is involved. This tendon must be surgically repaired to avoid telecanthus deformity

- In this setting, too, I have noticed providers ordering mandibular radiographs in lieu of CT scans. This is not acceptable if the patient can tolerate a CT scan.
- Open reduction internal fixation (ORIF) is required for fractures involving the parasymphyseal bone, angle of the mandible, or body.

2.12 Orbital Soft Tissue Injuries

2.12.1 Bottom Line Up Front

It is every physician's imperative to preserve life, limb, and eyesight, and 40% of monocular blindness is associated with trauma [69]. That's why I include a section devoted to soft tissue injuries in a land of fractures. As one of the cardinal senses, vision is uniquely vulnerable to trauma because of the eloquence of the parenchymal-neuro-vascular structures involved as well as their squishiness. In one retrospective review of polytrauma patients, just shy of one-quarter had orbital bony injuries, and about 10% had orbital soft tissue injuries [70].

Five Diagnostic Pearls for the Resident/Fellow on Call:

- Proper diagnostic assessment of the orbits is accomplished with dedicated imaging protocols, whether CT or MRI. Trauma series exams often initially consist of a noncontrast head CT and a noncontrast maxillofacial CT. Do not say "normal" when describing the orbits in this setting, even if they look normal. You open yourself up to medicolegal liability problems.
- There are many types of orbital and ocular hardware that can confound a trauma picture. For example, oculoplastic prostheses and Lester-Jones drainage tubes may be mistaken for foreign bodies. For a thorough review of postoperative orbits, to include their effects on MR imaging, I recommend the Reiter Radiographics article [71].
- Colobomas and staphylomas distort the normal geometry of the globe. Do not mistake them for globe ruptures.
- When you learn what a dilated superior ophthalmic vein looks like, you will want to call it on every scan. Keep in mind that there are benign reasons for this finding. It's not always a cavernous carotid fistula (CCF)!
- Conversely, in a patient with a recent history of trauma *and* presenting with chemosis, diplopia, and proptosis, raise the possibility of a CCF (Fig. 2.29).

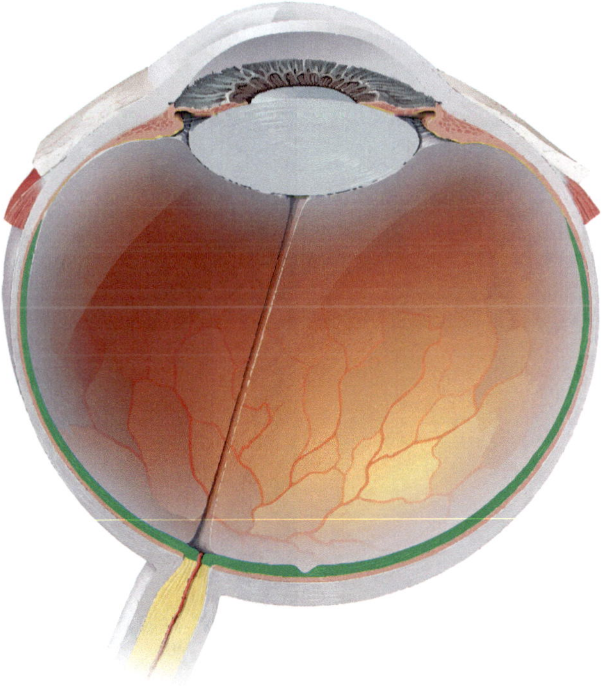

Fig. 2.29 Axial magnified illustration of the globe. The posterior green material is the territory of the retina. Notice that it stops at 10 o'clock and 2 o'clock positions of the ora serrota. **If the hemorrhage goes beyond the 10 and 2 positions, it is a choroidal hemorrhage**. Image courtesy of KenHub

- Hyphema
  - Normally, aqueous humor lives in the anterior chamber of the eye. If blood accumulates, it can block the patient's vision or the physician's assessment of the retina.
    - On CT, look for subtle effacement of the anterior chamber and hemorrhagic products.
    - Intraocular hypertension is a known complication of hyphema, especially higher grades [72].
- Lens Dislocation [Ectopia Lentis]
  - The lens of the eye can be subluxated from its attachments, hanging like a sheet on a washing line or settling in the dependent segment of the globe.
  - Occurs after disruption of the suspensory ligaments that connect the lens to the ciliary muscle.
  - Often a clinical diagnosis, but hyphema can obscure the clinical exam. CT in this instance is diagnostic (Fig. 2.30).

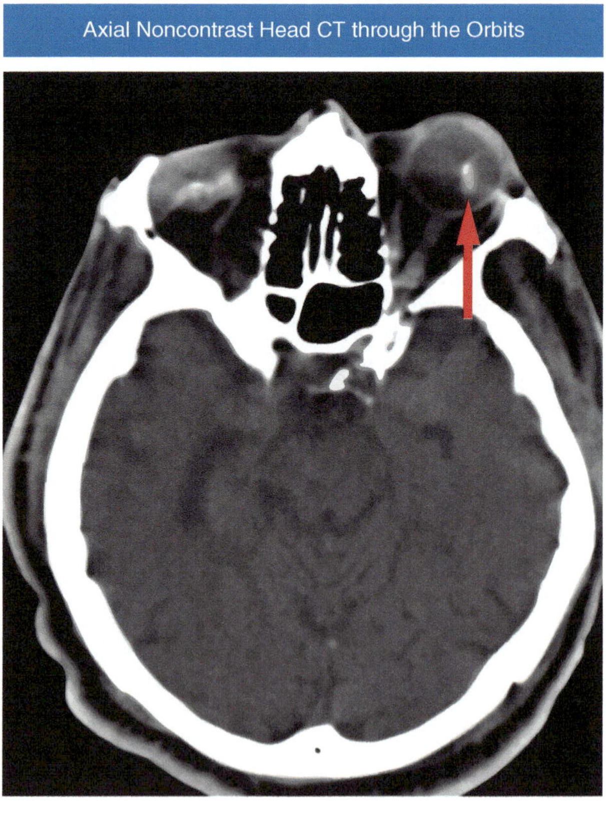

Fig. 2.30 An annotated CT image through the orbits reveals the patient's left lens flapping from its hinge into the globe. The patient was punched in the face in this instance; however, ectopia lentis can also be spontaneous in patients with Marfan's disease

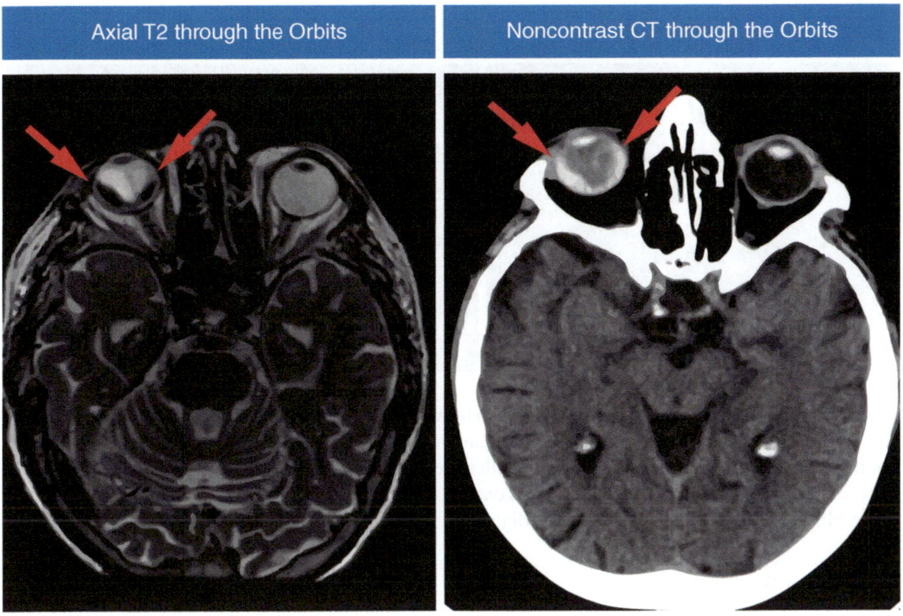

Fig. 2.31 The MRI on your left and the CT on your right are annotated at the 10 and 2 o'clock positions of the patient's right eye. Observe the lenticular disc-like structures laterally and medially that seem to converge at the optic disc at the back of the globe. This is consistent with retinal hemorrhage and detachment

- Choroidal Hemorrhage
  - On CT, choroidal hemorrhages can be seen as hemorrhagic products along the periphery of the globe that can be near circumferential.
- Retinal Hemorrhage/Detachment
  - The retina is limited in anterior extent by the ora serrata. Therefore, if you visualize a globe in the axial plane like a clock, 12 will be most anterior and 6 will be most posterior. A retinal hemorrhage will go no farther than 10 laterally and 2 medially.
  - In the axial CT plane, these will appear as the letter V with the vertex at the optic disc around 6 (Fig. 2.31).
- Retrobulbar Hematoma
  - A collection of blood behind the globe of sufficient size may anteriorly displace it, stretching the optic nerve. This may result in blindness (Fig. 2.32).
- Globe Rupture
  - Irregularity of the globe on CT is a troubling sign of globe rupture, which can result in blindness, endophthalmitis, sympathetic ophthalmia, and death [73].
    - Sympathetic Ophthalmia can occur after globe trauma. It is thought to be an autoimmune process in which the non-injured eye is attacked by autoantibodies generated in response to injury to the other.

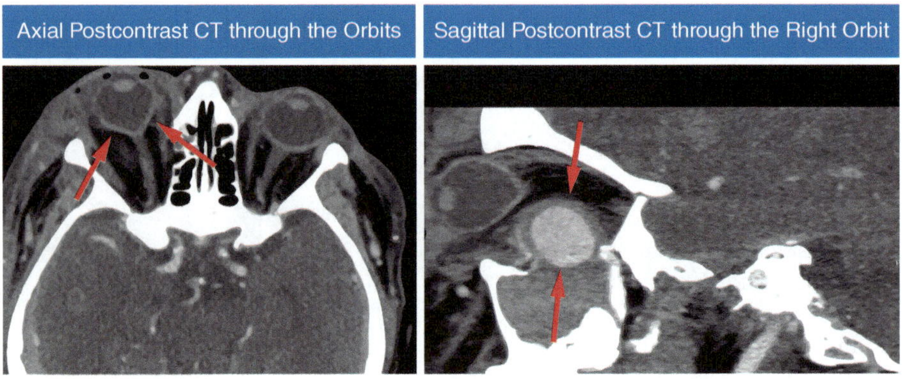

Fig. 2.32 The annotated left axial CT image shows suspicious straightening of the posterior margins of the right globe and tenting of the optic nerve insertion. This finding was the result of orbital trauma that allowed the growth of a retrobulbar hematoma, which is annotated on your right. We have an ophthalmologic emergency in this case, as such strenuous traction on the optic nerve can cause blindness. Call the ER team immediately!

- Foreign Bodies
  - Metallic foreign bodies will be bright on CT imaging and may generate beam hardening and/or streak artifact.
  - Metallic foreign bodies will generate susceptibility artifact on gradient and SWI sequences.
  - Graphite in pencils can be bright on CT.
  - Fresh (green) wood can be hypodense or isodense on CT.
  - I recommend *never* saying that a foreign body is not present on any exam. There are too many variables. Rather, say “No (CT or MRI) evidence of orbital foreign body.” Some of them can be imaging occult.

2.13 Temporal Bone Fractures

2.13.1 Bottom Line Up Front

The petrous temporal bone is the densest bone in the human body that requires approximately 1875 lbs. (850 kg) of force to fracture [74]. These injuries are unilateral in 80–90% of cases, thankfully, because the amount of anatomy that can be disrupted—and thus potential morbidity—is substantial [75]. In days gone:: old man voice::, t-bone fractures were classified as longitudinal, transverse, and mixed. Of course, additional schemes have come along as people want to get their names out there, but I find that just describing the anatomy in clear, concise, and practical terms is the best way to go about it. The exception here is whether or not the fracture line spares or involves the otic capsule. That’s quite important.

Five Diagnostic Pearls for the Resident/Fellow on Call:

- Evaluation of the temporal bones requires axial and coronal reformations in bone and soft tissue kernel at **1 mm slice thickness or less**. Routine CT head imaging cannot adequately assess the T-bones.
- When assessing a head trauma, look for secondary clues of temporal bone fracture to include mastoid effusion, pneumolabyrinth, and pneumocephalus [75, 76]. Otherwise, clinical assessment is essential in proper patient dispositioning.
- Fractures through the carotid canal indicate a 35% chance of internal carotid artery injury [77]. If you see this, you must recommend MR or CT angiography for further assessment. **Venography is similarly required if the jugular foramen or sigmoid sinus is fractured**.
- There are many sutures, foramina, canals, and articulations in this region that can mimic fractures. It really is one of the most difficult areas in the human body to assess with modern imaging. See below table [78] (Table 2.3).
- Do not get hung up in the details of a t-bone fracture when evaluating an acute trauma. Prioritize searching for findings that will kill, blind, or paralyze the patient in the immediate term.

Table 2.3 A list of the various foramina, fissures, and canals of the temporal bone. Some contain clinically important structures, while others can act as fracture mimics

Skull bone	Foramen/canal/ fissure	Neurovascular structures	Clinical implications
Temporal	Internal auditory canal	CN VII, VIII	Deafness, vertigo
Temporal	Carotid canal	Carotid artery	Carotid dissection/ pseudoaneurysm
Temporal	Jugular foramen (shared with occipital bone)	Jugular vein, CN IX, X, XI (pars vascularis gets X and XI; pars nervosa gets CNIX and inferior petrosal vein)	Venous thrombosis, CN palsies
Temporal	Foramen lacerum (shared with sphenoid bone)	Nothing	Carotid artery injury associated with petrous-lacerum junction fractures
Temporal	Stylomastoid foramen	CN VII	Facial paralysis
Temporal	Petrotympanic fissure (Glaserian canal)	Chorda tympani (VII), tympanic artery	Dysgeusia, xerostomia
Temporal	Vestibular aqueduct	Endolymphatic duct and sac	Deafness
Temporal	Cochlear aqueduct	Perilymphatic duct	–
Temporal	Petroclival fissure (petro-occipital)	Inferior petrosal sinus	–
Temporal	Petrosphenoidal fissure	–	–

1. Otic Capsule Sparing Temporal Bone Fracture

   Also known as the bony labyrinth, the otic capsule contains the fluid that makes hearing possible. It is comprised of the cochlea, vestibule, and semicircular canals. T-bone fractures that spare the otic capsule will often violate the squamous temporal bone, mastoids, middle ear cavity and ossicular chain, tegmen mastoideum, and tegmen tympani [77].

   - Otic capsule sparing injuries are **more commonly associated with conductive hearing loss** [79].
   - Scrutinize carefully the tegmen "roof" of the mastoids and tympani. Dehiscence of these structures may portend a CSF leak.

2. Otic Capsule Disrupting Temporal Bone Fracture

   Disruption of the bony labyrinth is bad, mmkay? It results in sensorineural hearing loss, and conductive hearing loss too, as well as possessing a higher incidence of facial nerve paralysis and intracranial complications [80].

   - Look for pneumolabyrinth as an indicator of violation of the otic capsule. This finding should precipitate a careful multiplanar search.
   - Carefully search for fracture involvement along the entire course of the facial nerve (Figs. 2.33, 2.34, and 2.35).

Recommendation: I realize the T-bones are incredibly complex and may present an overwhelming burden to an already-stressed resident or fellow on call. Remember, the imaging will come across most often as a noncontrast head CT

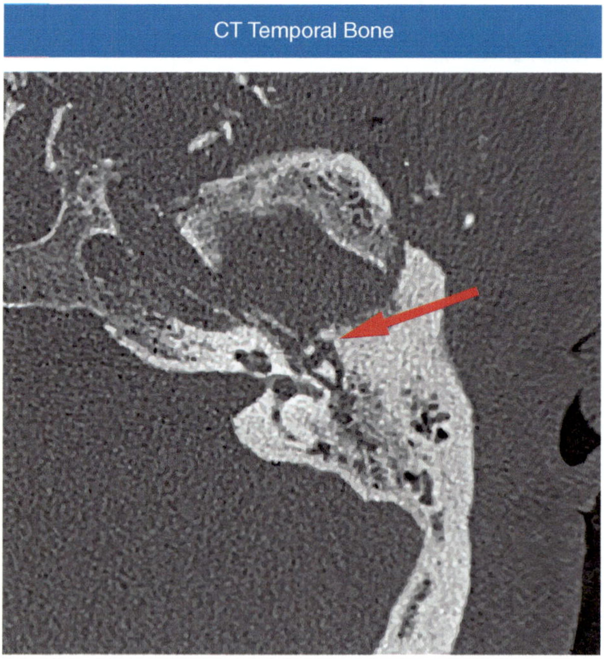

Fig. 2.33 Here we have an annotated axial thin-cut CT image through the left middle ear cavity. There is a transverse, otic capsule-sparing fracture of the left temporal bone. Notice the ice cream separated from the cone, consistent with incudomalleolar dislocation (arrow)

Fig. 2.34 Another annotated axial thin-cut CT image through the left temporal bone demonstrating a transverse fracture pattern that appears to spare the otic capsule

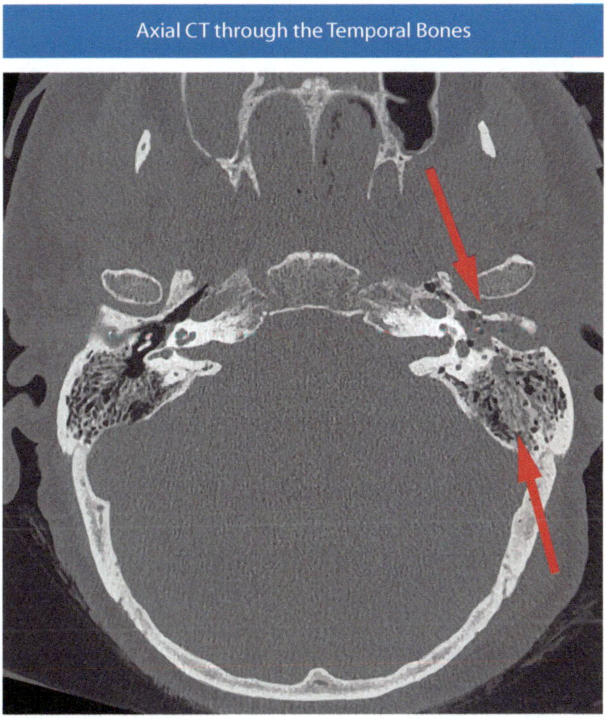

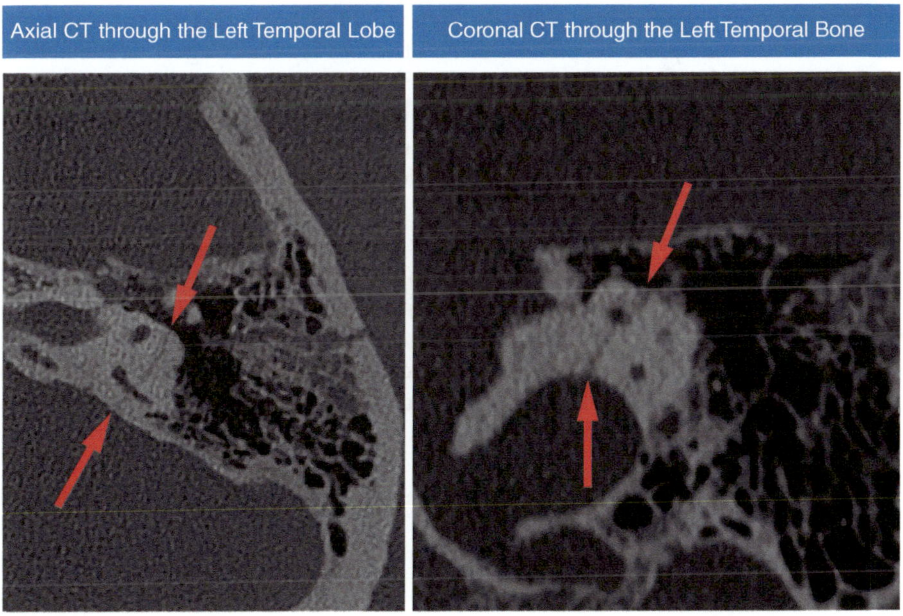

Fig. 2.35 Two annotated CT images through the left temporal bone reveal multiple left temporal bone fractures. Pay specific attention to the thin fracture line traversing the otic capsule and violating the semicircular canals. This is a pertinent finding that should be included in your report, including the impression, as these injuries can make a patient's life a living hell

and perhaps maxillofacial imaging. *Your* job is to find pathology that will kill or maim the patient in the immediate term. When evaluating the temporal bones on examinations not specifically tailored for them, think big.

- *Is there a unilateral mastoid effusion?*
- *Is there a fracture line that obviously violates the petrous temporal bone, and does it involve the otic capsule?*
- *Is there a fracture line that violates the carotid canal or jugular foramen?*

If and when the dedicated temporal bone scan comes across, elevate it to a senior resident or attending if you are not experienced enough to deal with it *or* you're caring for other critically ill patients.

2.14 Abusive Head Trauma (Non-accidental Trauma)

2.14.1 Bottom Line Up Front

Abusive head trauma is defined as intentional violent shaking and/or direct forceful blows to a child's head, most often children under the age of 2, and it is the most lethal form of child abuse [81]. A sad and unfortunate fact of practicing medicine is that you'll encounter this not infrequently, as it runs about 32–38 cases per 100,000 children in the West, and has increased significantly since 2012 [82].

Five Diagnostic Pearls for the Resident/Fellow on Call:

- This is a tricky diagnosis to make, as there will often be no specific outward signs of trauma in the pediatric patient, some of whom are under age 1 and unable to communicate.
- Noncontrast head CT will often be the first study ordered, and most (90%) of patients suffering from abusive head trauma will have subdural hemorrhage. However, subdural is not a slam dunk, demonstrating only moderate specificity [83].
- MRI of the brain is superior in assessing for DAI, searching for small hemorrhages, and dating them. Further, if abusive head trauma is confirmed, 75% of patients will have concomitant spinal injuries, and thus, MRI c-spine is indicated [83].
- **Abusive head trauma requires a multidisciplinary approach. Radiology is part of the equation. Primary care needs to corroborate the circumstances of the abuse. Ophthalmology will assess for retinal hemorrhages.** You get the picture.
- Gold standard definitional criteria for abusive head trauma and non-abusive head trauma do not exist [84] (Fig. 2.36).

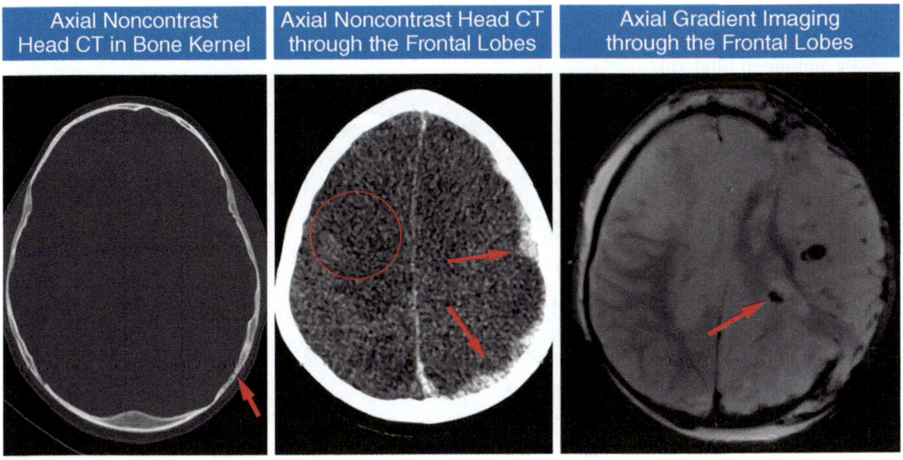

Fig. 2.36 The left CT image is annotated with an arrow revealing a subtle nondepressed left parietal skull fracture. The middle axial CT image demonstrates evidence of cytotoxic edema (circle) and a left subdural hematoma (arrows). The right MRI image demonstrates intraparenchymal hemorrhage (arrow) as well as left frontoparietal craniectomy. This was a fatal case of abusive head trauma

1. Subdural Hemorrhage

   Our old nemesis, subdural, has returned. However, it will be of significant benefit to discuss this pathologic process in the context of abusive head trauma because there are key differences *and* some interesting developments in recent literature.

   - Subdural hemorrhages, especially in children under 2 years, should raise suspicion for abusive head trauma. Full stop.
     - Evidence suggests that mixed-density subdural hematomas are not an automatic indicator of abusive head trauma [85, 86]
   - Subdural hemorrhages should not be automatically labeled as acute or chronic on CT but rather described by their density. Hygromas, hematohygromas, and regular subdurals can have overlapping and mixed densities.
   - MRI is better for qualifying the *minimum age* of a collection. If it has enhancing neomembranes visible on MRI, it's likely 2–4 weeks old [86].
   - One retrospective study more strongly associated subdural hemorrhage with abusive head trauma in cases where the victim presented with seizures, had multiple subdurals, or had subdurals >5 mm in thickness [87] (Fig. 2.37).
   - **There are a number of diseases that can predispose a young patient to subdural hemorrhage or the appearance of such: rickets, glutaric aciduria type 1, Menke's disease, bleeding diatheses, and benign enlargement of the subarachnoid spaces** [88].

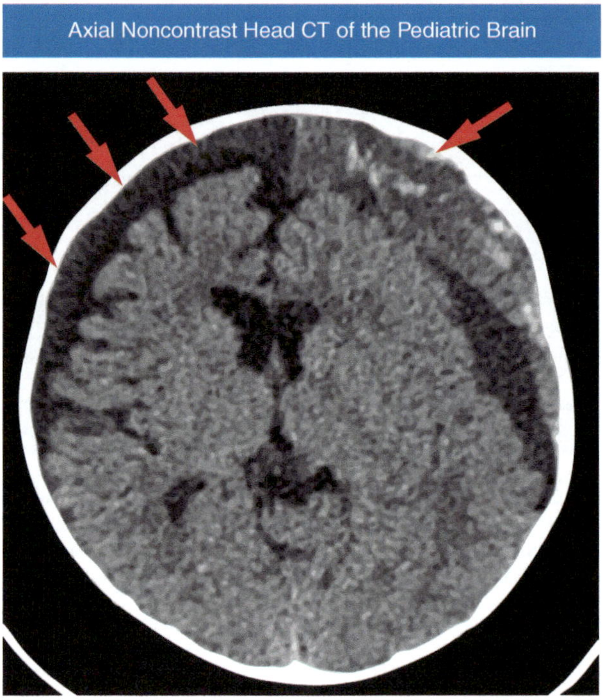

Fig. 2.37 An axial CT through the pediatric brain reveals annotated bilateral subdural hematomas. They are of mixed density, which suggests different injury chronologies. These findings are suspicious and should precipitate a workup for abusive head trauma

2. Skull Fracture

   Kids crack their domes rather frequently in day-to-day life—roughhousing, sports, bikes, or doing dumb things. However, there are some clues that may suggest abusive head trauma.

   - Fractures that span a segment of skull and contact two or more sutures were found to have a statistically significant association with abusive head trauma [89].
   - A single forceful blow on a single point of the calvarium can create multiple skull fractures, so multiple fractures are not pathognomonic.
   - The most commonly found fracture in abusive head trauma is a linear nondisplaced fracture of the parietal bone. Unfortunately, this is the most common fracture in accidents, too [90].
   - There are multiple fracture mimics lying in wait for the neophyte radiologist: wormian bones, vascular channels, anomalous sutures, regular sutures, and pyknodysostosis, to name a few [88].

Remember: Radiology is a consultant service. Do not accept pressure to use certain language that may indict someone for the sake of convenience. Abusive head trauma does have false positives. The best course of action in this scenario is to describe in the most objective terms and avoid ascribing a finding to a certain timing. Staff pediatric- and neuroradiologists should be where the buck stops in this instance, not the on-call resident/fellow.

2.15 Cervical Spine Trauma

The normal cervical spine has seven elements with the first two—C1 and C2—being named the atlas and the axis, respectively. It possesses a lordotic curve. There are uncinate processes, which help ensconce the intervertebral discs. The bilateral vertebral arteries pass through the foramina transversaria of the cervical spine on their way to the confluence. And significant trauma to the cervical spine puts the patient in a world of shit.

- Acute Fractures of the Cervical Spine
  - Jefferson Fracture
  - Hangman Fracture
  - Dens Fractures
  - Flexion Teardrop Fracture
  - Extension Teardrop Fracture
  - Dissociation and Subluxation Injuries
- Soft tissue Injuries/Hemorrhage
  - Pseudoaneurysm
  - Dissection
  - Spinal Epidural and Subdural Hematoma

After the noncontrast head CT and (probably) maxillofacial CT, you'll probably see the noncontrast cervical spine CT. In patients found down for significant trauma to the head, approximately 8% will have concomitant cervical spine injury [91]. Therefore, don't be surprised if these studies roll across in pairs (Fig. 2.38).

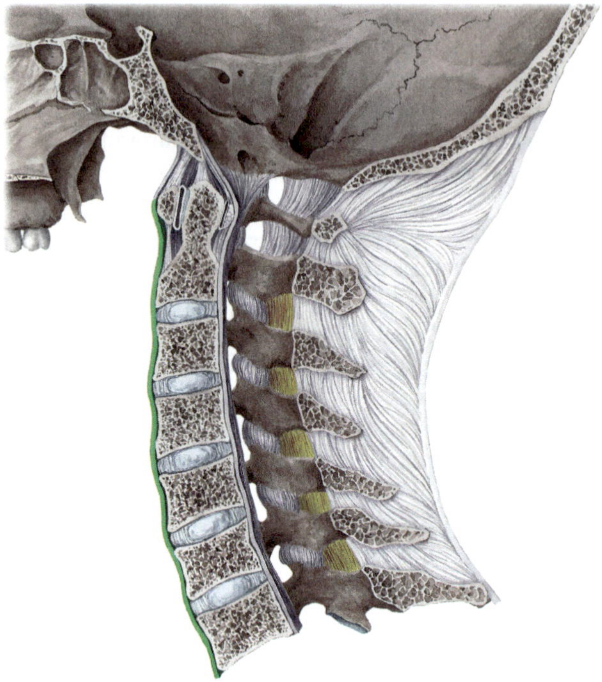

Fig. 2.38 Sagittal illustration of the cervical spine demonstrating a normal lordotic configuration. The green linear structure seen anteriorly is the anterior longitudinal ligament. At the posterior margin of the vertebral bodies, the white linear structure is the posterior longitudinal ligament. Image courtesy of KenHub

2.16 Acute Fractures of the Cervical Spine

2.16.1 Bottom Line Up Front

Fractures of the cervical spine are dependent upon the mechanism of injury, and thus, you should seek this information out. For example, axial loading injuries like diving into a shallow pool will burst vertebral bodies and rupture the C1 ring, whereas hyperflexion injuries like head-on collisions will cause flexion teardrop fractures.

Five Diagnostic Pearls for the Resident/Fellow on Call:

- Again, multiplanar CT soft tissue and bone kernel images are essential in making the proper diagnosis in this setting. Be aware that significant spinal cord injury may be present in the absence of discrete cervical spine fractures, the so-called SCIWORA. Therefore, MRI is not excluded from the diagnostic workup.
- There are a number of pitfalls when diagnosing possible cervical spine fractures, including the wily *limbus* vertebra, vascular channels, os odontoideum, and persistent ossiculum terminale, to name a few.
- **If a fracture line intercepts or crosses a transverse foramen, you must recommend CT angiography of the neck** [92].

- Disc material can occasionally be seen on CT imaging. Finding a traumatic herniation, or even a significant incidental herniation, can go a long way in helping your patient.
- Always check the thyroid parenchyma and the lung apices. I have found multiple thyroid carcinomas, lung cancers, and pneumothoraces while searching for fractures.

1. Jefferson Fracture

    These occur when axial (vertical) loading on the neck occurs that transmits enough force through the occipital condyles to the ring of C1. Thankfully, neurologic injuries from Jefferson Fractures are rare.

    - For those interested, the *Gehweiler and Jefferson* classification systems are used for these fractures (Fig. 2.39).
    - Descriptions of the injury are generally enough for neurosurgeons [93].
        - Injuries involving only the anterior arch are usually managed conservatively.
        - Injuries involving the anterior and posterior arches or the lateral masses may mean transverse ligament injury and instability, thus surgery.
    - **The transverse ligament courses behind the dens** and keeps it fixed in place and out of the spinal canal. The transverse ligament attaches to two tubercles on the C1 arch. Sometimes, bony tubercle avulsions can be seen.
    - Sometimes you may get an open-mouth dens radiograph. The lateral masses of C1 should be equidistant from the dens. If they're displaced symmetrically, query a possible Jefferson burst fracture. If one is displaced, a fracture is still possible, but also make sure the patient's head isn't rotated (Fig. 2.40).

2. Hangman Fracture

    Traumatic spondylolysis of C2, or bilateral C2 pars interarticularis defects, is the descriptive mechanism of injury for the vintage hangman fracture. More specifically, the injury is caused by rotational, distraction, and hyperextension forces applied to the craniocervical junction—hanging. A method of execution since time immemorial, a "humane" hanging requires the noose knot to be placed laterally on the victim's neck in order to effect the rotational forces necessary for the fracture-dislocation injury that yields instant death. Otherwise, it's not so instant. Most suicide attempts are not instant, with death resulting from asphyxiation.

    - Most modern-day hangman fractures occur due to high-speed automobile accidents, and they are more common in the elderly [94].
    - These fractures can be classified as typical or atypical.
        - A typical hangman fracture will show bilateral axis pars interarticularis defects and widening of the C1-C2 interspinous interval.

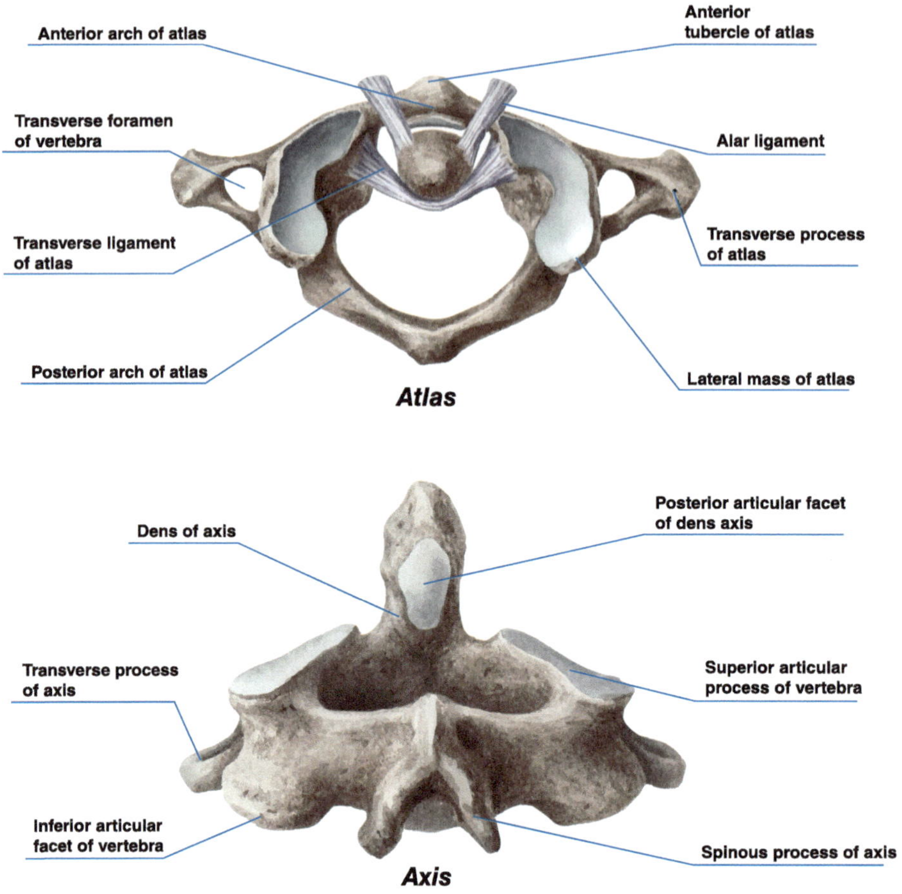

Fig. 2.39 AP and axial views of the atlas and axis. Notice the importance of the transverse ligament of the atlas. It keeps the dens secured in place and out of the spinal canal. Image courtesy of KenHub

  - An atypical hangman fracture can involve the posterior vertebral body of C2 as opposed to pars. These are known to have more complications with conservative management [95].
- There are a couple well-known classification systems for hangman fractures.
  - *Levine-Edwards*: three types
  - *Francis-Fielding*: five types

3. Odontoid (Dens) Fractures

   The most common fracture to involve the axis, the dens fracture, has a bimodal age distribution, and will become a staple in your day-to-day practice [96]. The usage of cervical spine radiographs is still commonplace when

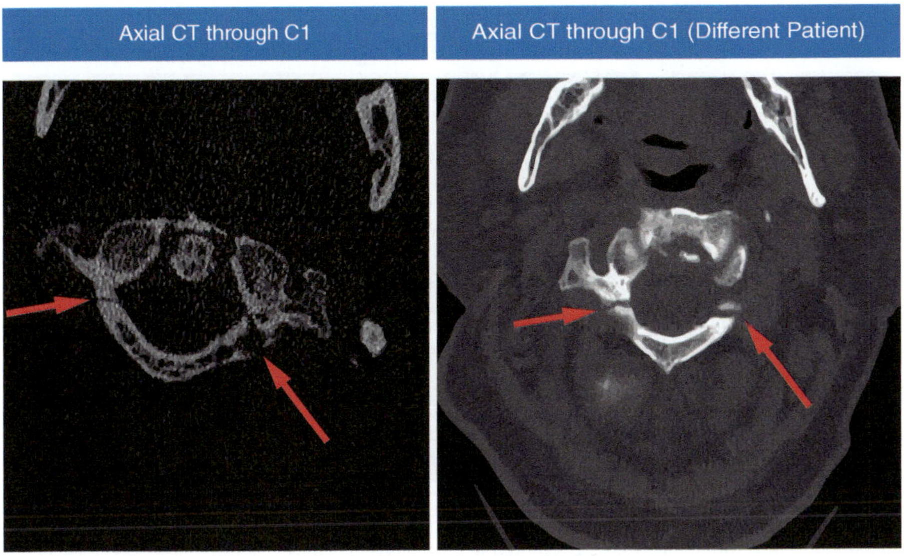

Fig. 2.40 Annotated axial CT images through the posterior arches of two atlases (C1) in two different patients. This injury pattern is often caused by axial loading. The image on your right shows some calcification of the transverse ligament of the atlas. **Do not mistake this for a tubercle fracture**

evaluating for fractures, but the sensitivity is quite low around 30% [97]. Therefore, recommend CT.

- In this instance, classification of dens fractures is important, as the primary classification system (*Anderson-D'Alonzo*) has become widely accepted. However, it is imperfect.
  - Type 1: Tip of the dens. Rare but stable
  - Type 2: Base of the dens. Unstable. Often surgical. High rate of nonunion
  - Type 3: Subjacent to the base of the dens into the lateral masses. Stable
- More recent literature suggests the feasibility of conservative management for Type 2 dens fractures based on imaging measurements [98, 99].
- Be aware that os odontoideum (OO) is a fracture mimic that is not entirely benign in its own right [100].
  - A round ossicle not contiguous with the dens, which looks hypoplastic.
  - Can affect chronic repetitive microtrauma of the spinal cord *or* smash it during a hyperflexion event, leading to serious neurologic sequela (Fig. 2.41).
- The persistent ossiculum terminale is another variant, benign compared to the ornery OO, which sits above the alar ligament. It may mimic a Type 1 dens fracture; otherwise, this little fella is nothing to worry about (Figs. 2.42 and 2.43).

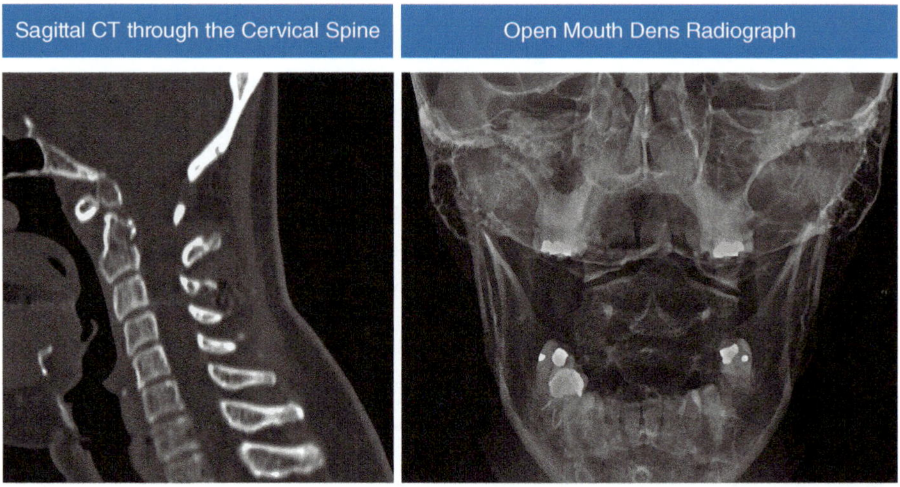

Fig. 2.41 Midsagittal CT (left) and open-mouth dens imaging (right) of the same patient who sustained neck trauma. One may be tempted to call a Type II Dens Fracture. However, look at the bony fragment. It's well-corticated. Look at the remainder of the dens. It's irregular. What we have here, people, is an os odontoideum. The open-mouth dens radiograph on the right shows an irregular lucency through the base of the dens. Nobody would fault you for calling an acute fracture on this radiograph. The CT is diagnostic, though

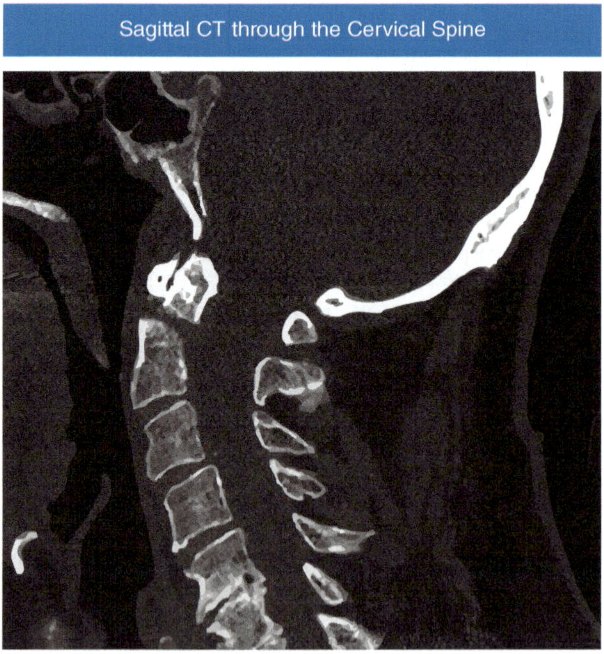

Fig. 2.42 Sagittal noncontrast CT through the mid cervical spine. There is an acute fracture through the base of the dens with posterior angulation and mild distraction. Note that there is a lack of cortication of the fracture fragments. Type II dens fractures were surgical lesions in the past, but some patients are now being conservatively managed with immobilization

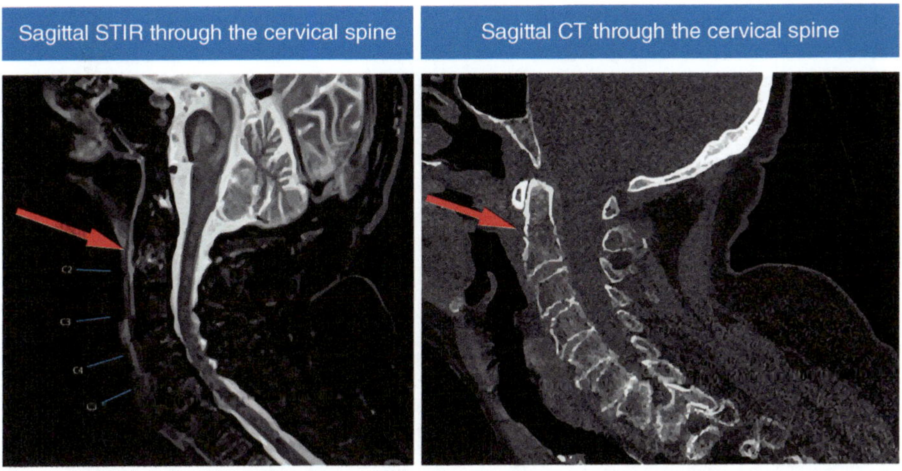

Fig. 2.43 On your left, we have an annotated sagittal STIR MRI image through the cervical spine. On the right, an annotated CT of the same patient. The arrows demonstrate a subtle fracture line that undermines the base of the dens, consistent with a type III dens fracture. The STIR sequence shows a little bit of edema. One would expect more edema in the setting of an acute fracture, right? **This is often not the case, and thus, a paucity of bone marrow edema should not dissuade you from calling an acute fracture**

4. Flexion Teardrop Fracture

   This fracture is recognized as the most severe type of fracture of the cervical spine. The mechanism is forceful hyperflexion with additional axial loading. The result will be triangular fractures of the anterior vertebral bodies, tearing of the posterior longitudinal ligament, retropulsion of vertebral body fragments into the spinal canal, and compromise of the spinal cord. Further, this injury pattern can also occur in the thoracic spine [101].

   - Remember the mechanism of injury when you ponder the injury patterns. Hyperflexion will impart extreme compressive forces on the anterior structures of the cervical column. Tensile forces will thus distract the posterior structures.
     - Look for injuries to the posterior longitudinal ligament, ligamentum flavum, interspinous ligament, zygapophyseal (facet) joints, and supraspinous ligament.
   - This is considered a highly unstable injury with a significant probability of spinal cord injury (Fig. 2.44).

5. Extension Teardrop Fracture

   Like its forward-vectored cousin, the extension teardrop fracture will result in anterior vertebral body endplate fractures, though the mechanism of injury is different. Whereas the flexion teardrop fracture is compressive force, the

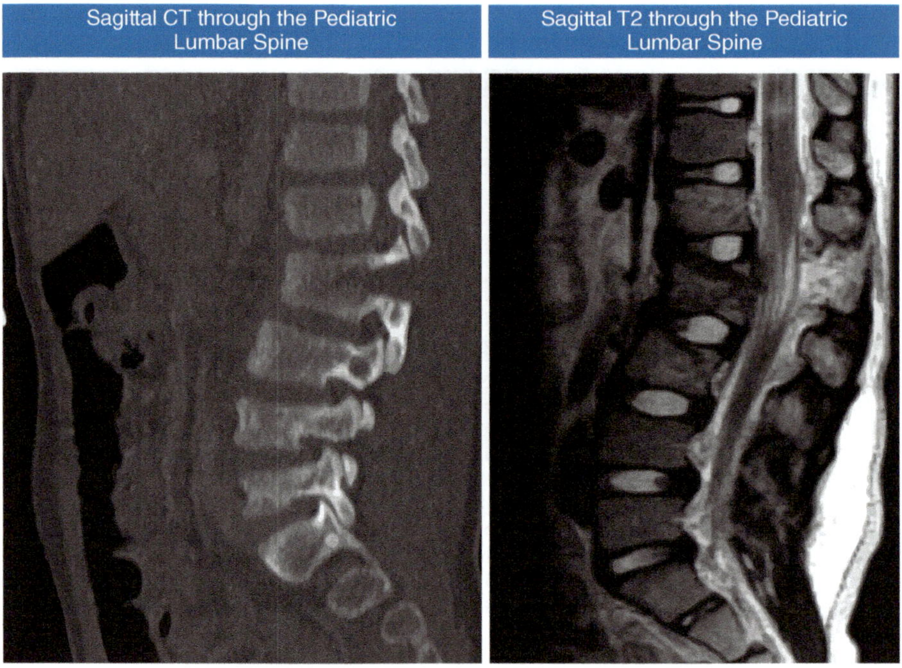

Fig. 2.44 Though not in the cervical spine, the mechanism of injury here is flexion, often over a lap seatbelt. The sagittal CT image reveals a horizontal fracture through the posterior L2 elements and vertebral body with diastasis. The subsequent MRI shows widening of the spinous processes and ligamentous injury. This is called a Chance fracture

extension teardrop fracture is due to tensile force applied across the anterior longitudinal ligament, resulting in ligamentous avulsion fractures.

- Most extension teardrop fractures will not result in neurologic injury.
- It is extremely uncommon to have extension teardrop fractures involve the axis, but it can happen.
- Look for tearing of the anterior longitudinal ligament.
- You may also see compression injuries of the facet joints or fractures of the spinous processes (known as clay-shoveler's fractures).

6. Dissociation and Subluxation Injuries

   Subluxation is a general term generally applied to vertebral bodies and other anatomy, slipped away from the place they're supposed to be. When taken in the context of the cervical spine, subluxation portends a very high risk of injury to the spinal cord, especially when trauma is the etiology. However, there are a multitude of other possible causes of cervical subluxation, including Down syndrome, rheumatoid arthritis, and Marfan syndrome, to name a few. Dissociation is another beast altogether.

   - Atlanto-Occipital Dissociation [102]
     - Also known as internal decapitation.

  - MRI can show tears through the alar ligaments, apical ligament, tectorial membrane, which is contiguous with the PLL, and the anterior atlanto-occipital membrane, which is contiguous with the inferior ALL.
  - There will be a distraction of the atlanto-occipital joint, which is formed by the superior articular facet of C1 and the occipital condyle.
  - Cerebrovascular injury is a concomitant finding.
  - The basion-dens interval of this injury will be increased to >10 mm.
  - It is frequently fatal.
- Perched Facets
  - Extreme flexion can cause the inferior articular process of one facet to "touch tips" with the superior articular process of its inferior neighbor.
- Jumped Facets
  - If the flexion is extreme enough, the inferior articular process of a facet can "jump" over the superior articular process of its inferior neighbor, essentially locking them in place (Fig. 2.45).

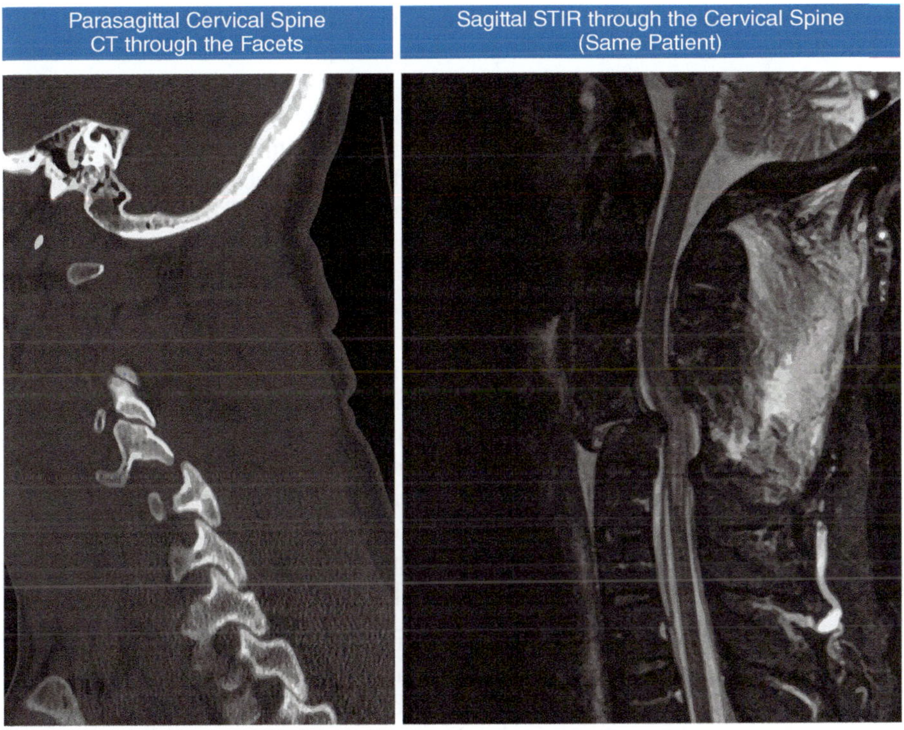

Fig. 2.45 The image on your left is a parasagittal CT through the cervical spine demonstrating a jumped facet. The inferior articular process of C5 has leapt over the superior articular process of C6. STIR MRI of the same patient's cervical spine reveals traumatic anterolisthesis of C5 on C6, traumatic posterior C5-C6 disc herniation that abuts the ventral spinal cord, spinal cord STIR signal abnormality, multiple tears of the posterior element ligaments, and prevertebral soft tissue thickening. Diagnosis? Hyperflexion injury after an automobile accident

2.17 Soft Tissue Injuries/Hemorrhage

2.17.1 Bottom Line Up Front

The neck is chock full of soft bits that are essential for not only being alive but also being *pleasantly* alive. Therefore, it comes as no surprise that trauma to the cervical spine and the surrounding soft tissues has significant morbidity and mortality. It is in this setting that an eye to fine detail can prevent an already sad and injured patient from experiencing further misery.

Five Diagnostic Pearls for the Resident/Fellow on Call

- About 12% of patients with a cervical spine fracture will have a concomitant spinal cord injury [103]. However, especially in adolescent patients, SCIWORA should always be considered. Remember, SCIWORA means spinal cord injury without radiographic abnormality.
- The prevertebral soft tissues are your friend and highly sensitive to cervical spine ligamentous injury, especially at C7 [104].
- Don't be afraid to obtain CT and MR angiograms in the setting of fracture. Both modalities have their shortcomings and can complement one another in making a diagnosis (more later).
- Scrutinize the cervical soft tissues in lung window. Subcutaneous and deep air may be indicative of tracheal fracture or esophageal injury.
- While evaluating CTs, make use of the soft tissue kernels. Look for evidence of traumatic disc herniations or spinal epidural hematomas. Follow up with MRI if possible.

1. Pseudoaneurysm

   By their very name, these things do not include each of the three layers of the artery: intima, media, and adventitia. Rather, they are vessel deformations/outpouchings caused by blood flow into an unnatural sub-intimal or sub-adventitial defect in the vessel, which can be spontaneous or traumatic. As the blood advances, it narrows the true lumen and forms a pouch contained only by the thin adventitia. This is a pseudoaneurysm. Cervical vessel pseudoaneurysms are often oval or round, possibly simulating mass lesions on MRI, and can also be mistaken for fibromuscular dysplasia [105].

   - Pseudoaneurysms are a common occurrence in the setting of military-type trauma (explosions) but also frequently associated with gunshot wounds [106].
   - Up to 30% of unstable cervical spine fractures will have concomitant vertebral artery injury, which can develop into pseudoaneurysm [107].
   - Carotid artery dissections will often evolve to pseudoaneurysms that are most frequently found in the subpetrous segment of the extracranial ICA [108].
   - Pseudoaneurysm formation can occur from hours to *years* after initial injury. The saccular formation can become a nidus for thrombus formation due to tepid blood flow, setting the patient up for stroke [109] (Fig. 2.46).

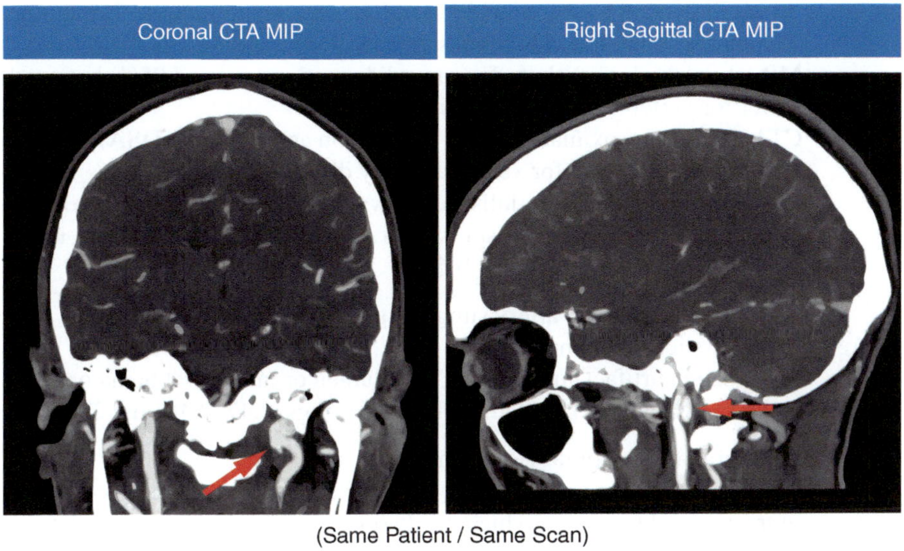

Fig. 2.46 Coronal (your left) and right parasagittal (your right) annotated CT angiogram MIP images. The patient has a left subpetrous pseudoaneurysm that projects medially (arrow). The same patient also has a RIGHT posterior subpetrous pseudoaneurysm (arrow). The pseudoaneurysm on the patient's right is basically undetectable on coronal imaging. Use those multiplanar reformats!

2. Dissections

When there is disruption of the intima of an artery, the flowing blood will dissect the intima away from the media, creating a false lumen.

- The vertebral arteries dissect quite easily, often with minor trauma such as chiropractic manipulation, sneezing, or sleeping with your head in a weird position. No, I'm not kidding [110, 111].
  - The most common presenting symptoms were occipital headache and stroke. Yes, a large number of vertebral dissections will go on to posterior fossa ischemia as we discussed in Chap. 1.
  - This disease more often afflicts young patients.
- Vertebral artery dissections can also occur secondary to significant trauma as well. Like gunshots.
- Spontaneous internal carotid artery dissection occurs more commonly than vertebral dissections but this process is a disease of mature patients [112].
  - Patients with spontaneous carotid artery dissection are much more likely to have subarachnoid hemorrhage and tinnitus.
- Carotid artery dissections can be either traumatic or spontaneous. Both seem to have positive outcomes with medical management [113].
- CT and MRI have good sensitivity and specificity for dissections and pseudoaneurysms:

- Axial T1 with fat saturation is useful in seeking intramural hematoma. Remember, fresh blood will be isointense!
- MRA has a high sensitivity for carotid dissections, around 95%, but much lower for vertebral artery dissections at 60%.
- CTA almost approximates digital subtraction angiography (DSA) in sensitivity when assessing for vertebral artery dissections [114].
- There are a number of pitfalls to be aware of when evaluating a possible dissection or pseudoaneurysm, which are excellently categorized in the AJR Provenzale et al. article [105]:
  - Fat surrounding an artery simulating hematoma
  - Entry slice phenomenon
  - Pulsation artifact simulating a false lumen or intramural hematoma
  - Ulcerated plaque simulating dissection (Figs. 2.47, 2.48, and 2.49)

3. Spinal Epidural and Subdural Hematoma

   The spinal dura mater has one layer, simply the meningeal layer, as compared to the intracranial dura mater, which has two layers. The spinal dura mater and the spinal arachnoid mater are in direct contact with one another, thus subdural hemorrhage in the spine, like in the head, occurs when the potential space is

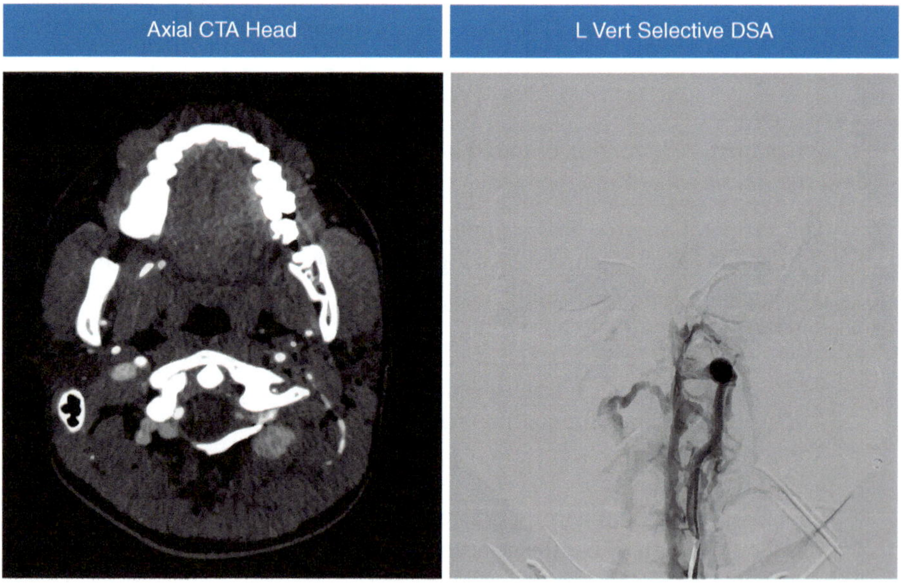

Fig. 2.47 Imaging of a patient who received a glancing bullet to the back of the head. The left image demonstrates an axial CTA through the level of C1. Notice the circular contrast opacification adjacent to the left posterior arch of C1. Active hemorrhage? Nope. The image on your right is the DSA after selection of the left vertebral artery. There is dense opacification of the left vertebral artery pseudoaneurysm, which has fistulized to the cervical venous plexus, and thus you also see early draining veins. The patient subsequently received coil embolization with complete resolution. Many thanks to Drs. Theresa Elder and Quang Nguyen for this case!

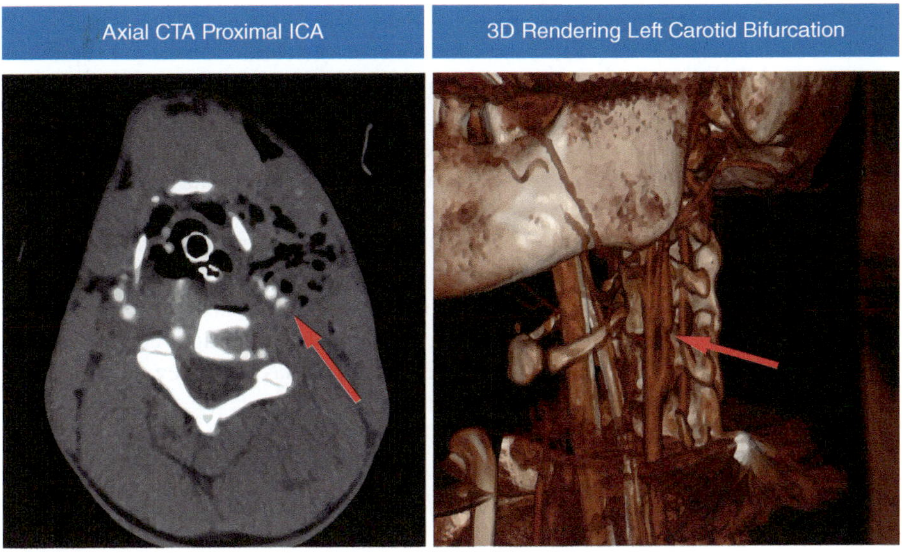

Fig. 2.48 Annotated axial CTA imaging (your left) through the proximal ICAs reveals an irregular filling defect in the cervical C1 segment of the ICA (arrow). Take note of the adjacent subcutaneous emphysema as well as the life support hardware midline. The patient was not having a good day. The traumatic dissection is also seen in the 3D rendering (arrow)

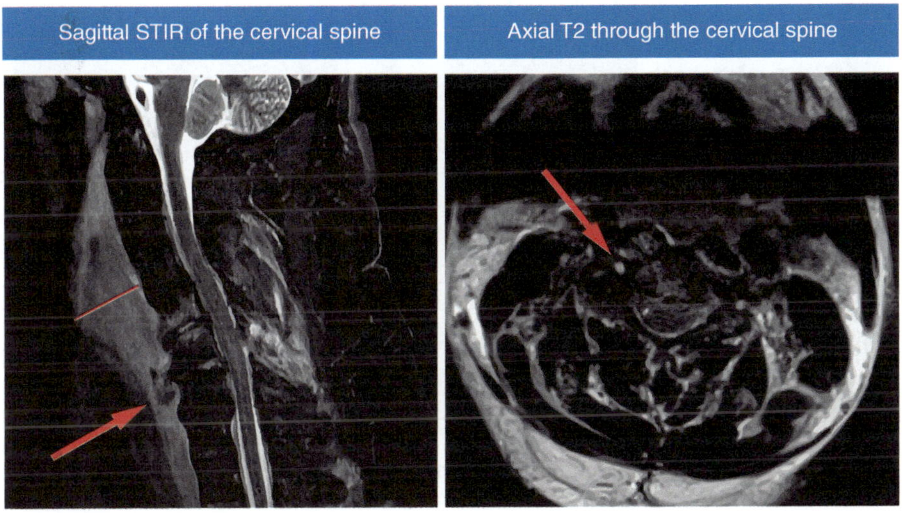

Fig. 2.49 This patient was involved in a high-speed automobile accident. The red line on the sagittal STIR image shows substantial prevertebral soft tissue thickening. The arrow on the same image shows tear of the anterior longitudinal ligament. There is additional cord signal abnormality at C5-C6. On your right, an arrow points to the right vertebral artery, which is lacking the normal flow void. This finding was proven to be a right vertebral artery dissection

filled with blood. Everything outside of the external margin of the thecal sac is therefore epidural. The fat is epidural. The vertebral bodies are epidural. Technically, Europa, the smallest of Jupiter's four Galilean moons is epidural if you want to get weird about it O_o.

- Spinal epidural hematomas can be spontaneous, traumatic, or related to minor trauma, and the epidural space is the most common spinal compartment to experience hemorrhage [115].
- When assessing for epidural hematomas look for obfuscation of the epidural fat and indentation of the thecal sac.
- Always look at your sagittal T1 sequences when evaluating a spine because CSF signal in the thecal sac can sometimes obscure T2 bright epidural blood.
- Subdural hematoma in the spine is very rare and has some key differences:
  - The outer dura mater layer will not be deformed.
  - The epidural fat will be preserved
- Subdural hematoma in the spine can be tricky if it is late subacute—why?
  - Remember, BaBy, it will appear bright on T1 and T2 sequencing. What else appears bright on T1 and T2? Fat (Figs. 2.50 and 2.51).
    - **Case reports of subdural hematoma mimicking epidural lipomatosis are out there** [116].

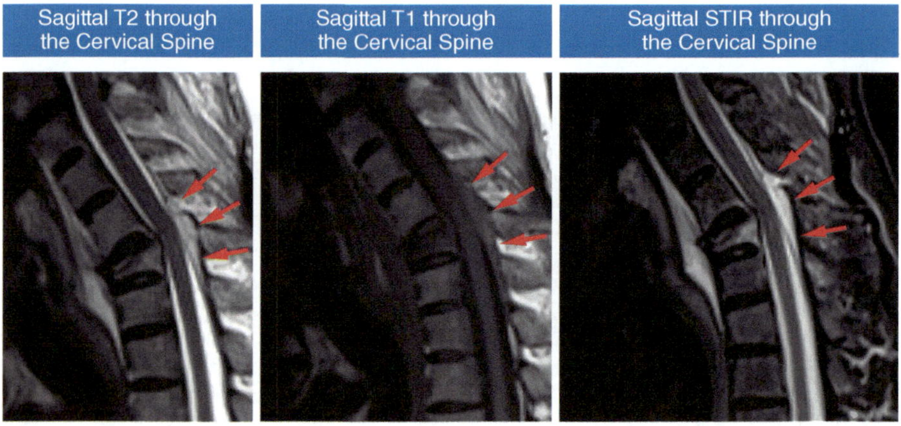

Fig. 2.50 Multisequence annotated imaging of the cervical spine. The patient sustained a C6 compression fracture and posterior ligamentous injury. Observe the dorsal discrete fluid collection effacing the dorsal epidural space. This was an acute cervical spinal epidural hematoma and they are quite easy to walk by

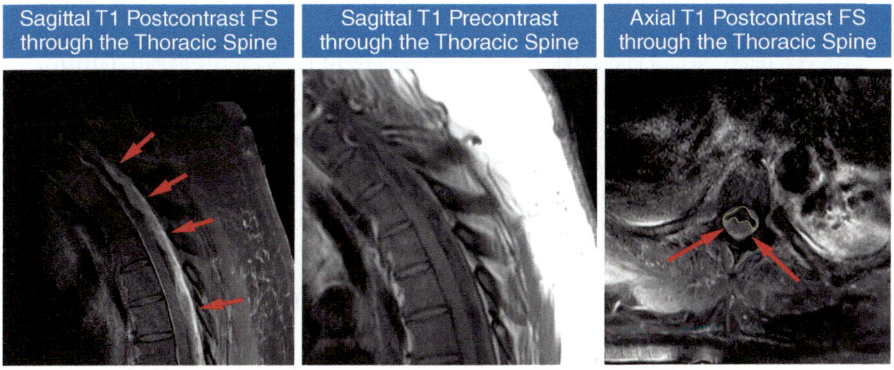

Fig. 2.51 MRI of the thoracic spine was obtained in a patient with bilateral upper extremity weakness. The left annotated sagittal T1 postcontrast fat-saturated image shows an enhancing dorsal epidural collection spanning multiple vertebral levels. The precontrast sagittal image does not show intrinsic T1 hyperintensity. The annotated axial T1 postcontrast fat saturated image on your right shows a large dorsal enhancing collection that partially effaces the thecal sac (circumscribed in yellow). This was a massive epidural abscess/phlegmon

References

1. Rahimi A, Corley JA, Ammar A, Shlobin NA, Rolle M, Mekary RA, et al. The unmet global burden of cranial epidural hematomas: a systematic review and meta-analysis. Clin Neurol Neurosurg. 2022;219:107313. https://doi.org/10.1016/j.clineuro.2022.107313.
2. Parker SL, Kabani AA, Conner CR, Choi PA, Withrow JS, Cai C, et al. Management of Venous Sinus – related epidural hematomas. World Neurosurg. 2020; https://doi.org/10.1016/j.wneu.2020.02.089.
3. ScienceDirect Topics. Epidural hematoma – an overview. Crit Care Clin. 2016;32(1) https://doi.org/10.1016/j.ccc.2015.08.009.
4. Scafa AK, Jiang T, Pescatori L, Corsini M, Piccirilli M. Association between spontaneous intracranial epidural hematoma and craniofacial infections: a systematic literature review. Surg Neurol Int. 2023:14. https://doi.org/10.25259/SNI_1068_2022.
5. Babatola BO, Salman YA, Abiola AM, Okezie KO, Oladele AS. Spontaneous epidural Haematoma in sickle cell Anaemia: case report and literature review. J Surg Tech Case Report. 2012;4(2) https://doi.org/10.4103/2006-8808.110271.
6. Bisen YT, Korde P, Dighe O, Iratwar S, Bisen G. Decompressive craniectomy in the management of low Glasgow coma score patients with extradural hematoma: a review of literature and guidelines. Cureus. 2023;15(1):e33790. https://doi.org/10.7759/cureus.33790.
7. Heino I, Sajanti A, Lyne SB, Frantzén J, Girard R, Cao Y, et al. Outcome and survival of surgically treated acute subdural hematomas and postcraniotomy hematomas – a retrospective cohort study. Brain Spine. 2023;3:102714. https://doi.org/10.1016/j.bas.2023.102714.
8. Karibe H, Narisawa A, Nagai A, Yamanouchi S, Kameyama M, Nakagawa A, et al. Incidence of chronic subdural hematoma after mild Head trauma in elderly patients with or without pre-traumatic conditioning of anti-thrombotic drugs. Neurol Med Chir (Tokyo). 2023;63(3):91–6. https://doi.org/10.2176/jns-nmc.2022-0327.
9. Grelat M, Madkouri R, Bousquet O. Acute isodense subdural hematoma on computed tomography scan—diagnostic and therapeutic trap: a case report. J Med Case Rep. 2016;10(1) https://doi.org/10.1186/s13256-016-0822-x.

10. Boujemâa H, Góngora-Rivera F, Barragán-Campos H, Karachi K, Chiras J, Sourour N. Bilateral acute subdural hematoma from ruptured posterior communicating artery aneurysm: a case report. Interv Neuroradiol. 2006;12(1) https://doi.org/10.1177/159101990601200107.
11. Nouri A, Gondar R, Schaller K, Meling T. Chronic subdural hematoma (cSDH): a review of the current state of the art. Brain and Spine. 2021; https://doi.org/10.1016/j.bas.2021.100300.
12. Yadav YR, Parihar V, Namdev H, Bajaj J. Chronic subdural hematoma. Asian. J Neurosurg. 2016;11(4) https://doi.org/10.4103/1793-5482.145102.
13. Edlmann E, Giorgi-Coll S, Whitfield PC, Carpenter KLH, Hutchinson PJ, Edlmann E, et al. Pathophysiology of chronic subdural haematoma: inflammation, angiogenesis and implications for pharmacotherapy. J Neuroinflammation. 2017;14(1) https://doi.org/10.1186/s12974-017-0881-y.
14. Hasan D, Nikoubashman O, Pjontek R, Stockero A, Hamou HA, Wiesmann M. MRI appearance of chronic subdural hematoma. Front Neurol. 2022:13. https://doi.org/10.3389/fneur.2022.872664.
15. Yu J, Tang J, Chen M, Ren Q, He J, Tang M, et al. Traumatic subdural hygroma and chronic subdural hematoma: a systematic review and meta-analysis. J Clin Neurosci. 2023:107. https://doi.org/10.1016/j.jocn.2022.11.010.
16. Elghamudi TA, Khan MN. Differentiating acute-on-chronic subdural hematoma and acute subdural Hematohygroma | Canadian journal of neurological sciences | Cambridge Core. Can J Neurol Sci. 2023;50(1) https://doi.org/10.1017/cjn.2021.240.
17. Subdural hematohygroma | Radiology Reference Article | Radiopaedia.org. doi: 10.53347/rID-160500.
18. Shoaib A, Hussain F, Khan M, Sohail A, Panjwani MH, Ashraf MT, et al. Comparative efficacy of craniotomy versus craniectomy in surgical management of acute subdural hematoma: a systematic review and meta-analysis. J Clin Neurosci. 2024:124. https://doi.org/10.1016/j.jocn.2024.04.010.
19. Rau CS, Wu SC, Hsu SY, Liu HT, Huang CY, Hsieh TM, Chou SE, Su WT, Liu YW, Hsieh CH, et al. Concurrent types of intracranial hemorrhage are associated with a higher mortality rate in adult patients with traumatic subarachnoid hemorrhage: a cross-sectional retrospective study. Int J Environ Res Public Health. 2019;16(23) https://doi.org/10.3390/ijerph16234787.
20. Kim J, Lee SJ. Traumatic subarachnoid hemorrhage resulting from posterior communicating artery rupture. Int Med Case Rep J. 2020:13. https://doi.org/10.2147/IMCRJ.S254160.
21. Sanicola HW, Stewart CE, Luther P, Yabut K, Guthikonda B, Jordan JD, et al. Pathophysiology, management, and therapeutics in subarachnoid hemorrhage and delayed cerebral ischemia: an overview. Pathophysiology. 2023;30(3) https://doi.org/10.3390/pathophysiology30030032.
22. Chen S, Luo J, Reis C, Manaenko A, Zhang J. Hydrocephalus after subarachnoid hemorrhage: pathophysiology, diagnosis, and treatment. Biomed Res Int. 2017;2017 https://doi.org/10.1155/2017/8584753.
23. Adams H, Ban VS, Leinonen V, Aoun SG, Huttunen J, Saavalainen T, et al. Risk of shunting after aneurysmal subarachnoid hemorrhage. Stroke. 2016;47(10) https://doi.org/10.1161/STROKEAHA.116.013739.
24. Vale FL, Bradley EL, Fisher WS. The relationship of subarachnoid hemorrhage and the need for postoperative shunting. J Neurosurg. 1997;86(3) https://doi.org/10.3171/jns.1997.86.3.0462.
25. Phan CM, Yoo AJ, Hirsch JA, Nogueira RG, Gupta R. Differentiation of hemorrhage from iodinated contrast in different intracranial compartments using dual-energy Head CT. Am J Neuroradiol. 2012;33(6) https://doi.org/10.3174/ajnr.A2909.
26. Bigler ED. Anterior and middle cranial fossa in traumatic brain injury: relevant neuroanatomy and neuropathology in the study of neuropsychological outcome. Neuropsychology. 2007;21(5):515.
27. Araki T, Yokota H, Morita A. Pediatric traumatic brain injury: characteristic features, diagnosis, and management. Neurol Med Chir. 2017;57(2) https://doi.org/10.2176/nmc.ra.2016-0191.

28. Quiñones-Ossa GA, et al. The puzzle of spontaneous versus traumatic intracranial hemorrhages. Egypt J Neurosurg. 2020;35:1. https://doi.org/10.1186/s41984-020-00084-9.
29. Tran C, Frederick H, Baqai H, Fairchild M, Solomon J, Aligabi A, et al. Traumatic brain injury and intraparenchymal hemorrhage progression: blood pressure variability matters. Am J Emerg Med. 2022;52:119–27. https://doi.org/10.1016/j.ajem.2021.12.005.
30. Kellner CP, Schupper AJ, Mocco J. Surgical evacuation of intracerebral hemorrhage. Stroke. 2021;52:10. https://doi.org/10.1161/STROKEAHA.121.032238.
31. Vo DT, Phan CC, Le HGN, Vo TP, Mai UTT, Le HK, et al. Diffuse axonal injury: a case report and MRI findings. Radiol Case Rep. 2022;17(1) https://doi.org/10.1016/j.radcr.2021.10.006.
32. Vieira RCA, Paiva WS, de Oliveira DV, Teixeira MJ, de Andrade AF, Sousa RMC. Frontiers | diffuse axonal injury: epidemiology, outcome and associated risk factors. Front Neurol. 2016:7. https://doi.org/10.3389/fneur.2016.00178.
33. Moenninghoff C, Kraff O, Maderwald S, Umutlu L, Theysohn JM, Ringelstein A, et al. Diffuse axonal injury at ultra-high Field MRI. PLoS One. 2015;10(3) https://doi.org/10.1371/journal.pone.0122329.
34. Park SE, Choi DS, Shin HS, Baek HJ, Choi HC, Kim JE, et al. Splenial lesions of the corpus callosum: disease Spectrum and MRI findings. Korean J Radiol. 2017;18(4) https://doi.org/10.3348/kjr.2017.18.4.710.
35. Ravikanth R, Majumdar P. Prognostic significance of magnetic resonance imaging in detecting diffuse axonal injuries: analysis of outcomes and review of literature. Neurol India. 2022;70(6) https://doi.org/10.4103/0028-3886.364066.
36. Jolly AE, Bălăeţ M, Azor A, Friedland D, Sandrone S, Graham NSN, et al. Detecting axonal injury in individual patients after traumatic brain injury. Brain. 2021;144(1) https://doi.org/10.1093/brain/awaa372.
37. Hamdeh SA, Marklund N, Lannsjö M, Howells T, Raininko R, Wikström J, et al. Extended anatomical grading in diffuse axonal injury using MRI: hemorrhagic lesions in the substantia Nigra and mesencephalic tegmentum indicate poor long-term outcome. J Neurotrauma. 2017;34(2) https://doi.org/10.1089/neu.2016.4426.
38. Jang SH, Kim SH, Chang MC. Injury of the arcuate fasciculus in the nondominant hemisphere by Subfalcine herniation in patients with intracerebral hemorrhage: two case reports and literature review. J Korean Neurosurg Soc. 2016;59(3) https://doi.org/10.3340/jkns.2016.59.3.306.
39. Santander X, Hidalgo YG, Flores JC, Gómez-Jordana B. Sinking skin syndrome in a decompressive craniectomy series: clinical and radiological features. Surg Neurol Int. 2022:13. https://doi.org/10.25259/SNI_582_2022.
40. Chawla H, Malhotra R, Yadav RK, Griwan MS, Paliwal PK, Aggarwal AD. Diagnostic utility of conventional radiography in Head injury. J Clin Diagn Res. 2015;9(6) https://doi.org/10.7860/JCDR/2015/13842.6133.
41. Assoumane II, Agada KN, Yahaya S, Hamma OI, Abdoulwahabou AM, Souleymane MA, et al. Surgical management and outcome of depressed skull fractures at Niamey National Hospital: about 233 cases. Interdisciplinary. Neurosurgery. 2024:36. https://doi.org/10.1016/j.inat.2023.101876.
42. Hamrah H, Mehrvarz S, Mirghassemi AM. The frequency of brain CT-scan findings in patients with scalp lacerations following mild traumatic brain injury; a cross-sectional study. Bull Emergen Trauma. 2018;6(1) https://doi.org/10.29252/beat-060108.
43. Aguas J, Conde C, Fructuoso GG, Mondelo FJ, Ferrer E. Giant cephalhematoma in a 15-year-old boy: unilateral amaurosis as the main complication. Surg Neurol. 1995;43(4) https://doi.org/10.1016/0090-3019(95)80065-O.
44. Bullock MR, Chesnut R, Ghajar J, Gordon D, Hartl R, Newell DW, et al. Surgical Management of Depressed Cranial Fractures. Neurosurgery. 2006;58(3) https://doi.org/10.1227/01.NEU.0000210367.14043.0E.
45. Jing XL, Luce E. Frontal sinus fractures: management and complications. Craniomaxillofac Trauma Reconstr. 2019;12(3) https://doi.org/10.1055/s-0038-1675560.

46. Wang J, Han F, Zhao Q, Xia B, Dai J, Wang Q, et al. Clinicopathological characteristics of traumatic head injury in juvenile, middle-aged and elderly individuals. Medical science monitor. Int Med J Exp Clin Rese. 2018:24. https://doi.org/10.12659/MSM.908728.
47. Rischall MA, Boegel KH, Palmer CS, Knoll B, McKinney AM, Rischall MA, et al. MDCT Venographic patterns of Dural venous sinus compromise after acute skull fracture. Am J Roentgenol. 2016;207(4) https://doi.org/10.2214/AJR.15.15972.
48. Bokhari R, You E, Bakhaidar M, Bajunaid K, Lasry O, Zeiler FA, Marcoux J, Baeesa S, et al. Dural venous sinus thrombosis in patients presenting with blunt traumatic brain injuries and skull fractures: a systematic review and meta-analysis. World Neurosurg. 2020:142. https://doi.org/10.1016/j.wneu.2020.06.117.
49. Hersh DS, Shimony N, Groves ML, Tuite GF, Jallo GI, Liu A, Garzon-Muvdi T, et al. Pediatric cerebral venous sinus thrombosis or compression in the setting of skull fractures from blunt head trauma. J Neurosurg Pediatr. 2018;21(3) https://doi.org/10.3171/2017.9.PEDS17311.
50. Sim SY, Kim HG, Yoon SH, Choi JW, Cho SM, Choi MS. Reappraisal of pediatric Diastatic skull fractures in the 3-dimensional CT era: clinical characteristics and comparison of diagnostic accuracy of simple skull X-ray, 2-dimensional CT, and 3-dimensional CT. World Neurosurg. 2017;108 https://doi.org/10.1016/j.wneu.2017.08.107.
51. Arora S, Vani K, Arora S, Vani K. Unusual pattern of leptomeningeal cyst with herniation of porencephalic cyst. Egypt J Radiol Nucl Med. 2023;54(1) https://doi.org/10.1186/s43055-023-01068-z.
52. Zhao Q, Ying J, Chen Y, Chen F, Zhang T, Jing J. Clinical and imaging characteristics of growing skull fractures in children. Sci Rep. 2024;14(1) https://doi.org/10.1038/s41598-024-56445-z.
53. Dreizin D, Sakai O, Champ K, Gandhi D, Aarabi B, Nam AJ, et al. CT of Skull Base fractures: classification systems, complications, and management. Radiographics. 2021;41(3) https://doi.org/10.1148/rg.2021200189.
54. Badakere A, Patil-Chhablani P. Orbital apex syndrome: a review. Eye Brain. 2019;11:63–72. https://doi.org/10.2147/EB.S180190.
55. Roselló EG, Granado AMQ, Garcia MA, Martí SJ, Sala GL, Mármol BB, et al. Facial fractures: classification and highlights for a useful report. Insights Imaging. 2020;11(1) https://doi.org/10.1186/s13244-020-00847-w.
56. Winegar BA, Murillo H, Tantiwongkosi B. Spectrum of critical imaging findings in complex facial skeletal trauma. Radio Graphics. 2013;33(1) https://doi.org/10.1148/rg.331125080.
57. Reiter MJ, Schwope RB, Theler JM, Reiter MJ, Schwope RB, Theler JM. Postoperative CT of the Midfacial skeleton after trauma: review of Normal appearances and common complications. Am J Roentgenol. 2017;209(4) https://doi.org/10.2214/AJR.17.17875.
58. Rhea JT, Novelline RA, Rhea JT, Novelline RA. How to simplify the CT diagnosis of Le Fort fractures. Am J Roentgenol. 2012;184(5) https://doi.org/10.2214/ajr.184.5.01841700.
59. Evans HH, Wurth BA, Penna KJ. Superior orbital fissure syndrome: a case report. Craniomaxillofac Trauma Reconstr. 2012;5(2) https://doi.org/10.1055/s-0032-1313363.
60. Yamsani B, Gaddipati R, Vura N, Ramisetti S, Yamsani R. Zygomaticomaxillary complex fractures: a review of 101 cases. J Maxillofac Oral Surg. 2015;15(4) https://doi.org/10.1007/s12663-015-0851-9.
61. Panneerselvam E, Ravi P, Sasikala B. Fractures of the Zygomaticomaxillary complex. In: Oral and maxillofacial surgery for the clinician; 2021. https://doi.org/10.1007/978-981-15-1346-6_56.
62. Connon FV, Austin SJB, Nastri AL. Orbital roof fractures: a clinically based classification and treatment algorithm. Craniomaxillofac Trauma Reconstr. 2014;8(3) https://doi.org/10.1055/s-0034-1393728.
63. Goyal P, Lee S, Gupta N, Kumar Y, Mangla M, Hooda K, et al. Orbital apex disorders: imaging findings and management. Neuroradiol J. 2018;31(2) https://doi.org/10.1177/1971400917740361.

64. Goodmaker C, Homan M, De Jesus O. NasNaso-orbito-ethmoid fractures. In:Treasure Island, FL: StatPearls Publishing. 2023. https://www.ncbi.nlm.nih.gov/books/NBK557468/. Accessed 18 Jan 2025.
65. Wei J-J, Tang Z-L, Liu L, Liao X-J, Yu Y-B, Jing W. The management of naso-orbital-ethmoid (NOE) fractures. Chin J Traumatol. 2015;18(5) https://doi.org/10.1016/j.cjtee.2015.07.006.
66. Epstein S, Ettinger RE. Nasal and Naso-orbito-ethmoid fractures. Semin Plast Surg. 2021;35(4) https://doi.org/10.1055/s-0041-1735791.
67. Naeem A, Gemal H, Reed D. Imaging in traumatic mandibular fractures. Quant Imaging Med Surg. 2017;7(4) https://doi.org/10.21037/qims.2017.08.06.
68. Nardi C, Vignoli C, Pietragalla M, Tonelli P, Calistri L, Franchi L, et al. Imaging of mandibular fractures: a pictorial review. Insights Imaging. 2020;11(1) https://doi.org/10.1186/s13244-020-0837-0.
69. Lin KY, Ngai P, Echegoyen JC, Tao JP. Imaging in orbital trauma. Saudi J Ophthalmol. 2012;26(4) https://doi.org/10.1016/j.sjopt.2012.08.002.
70. Goelz L, Syperek A, Heske S, Mutze S, Hosten N, Kirsch M, et al. Retrospective cohort study of frequency and patterns of orbital injuries on whole-body CT with maxillofacial multi-slice CT. Tomography. 2021;7(3):373–86. https://doi.org/10.3390/tomography7030033.
71. Reiter MJ, Schwope RB, Kini JA, York GE, Suhr AW. Postoperative imaging of the orbital contents. Radiographics. 2015;35(1) https://doi.org/10.1148/rg.351140008.
72. Gad K, Singman EL, Nadgir RN, Yousem DM, Pillai JJ, Gad K, et al. CT in the evaluation of acute injuries of the anterior eye segment. Am J Roentgenol. 2017;209(6) https://doi.org/10.2214/AJR.17.18279.
73. Yuan W-H, Hsu H-C, Cheng H-C, Guo W-Y, Teng MM-H, Chen S-J, et al. CT of globe rupture: analysis and frequency of findings. Am J Roentgenol. 2014;202(5) https://doi.org/10.2214/AJR.13.11010.
74. Patel A, Lofgren DH, Varacallo MA. Temporal Fracture. Treasure Island: StatPearls Publishing; 2024.
75. Kennedy TA, Avey GD, Gentry LR. Imaging of temporal bone trauma. Neuroimaging Clin N Am. 2014;24(3):467–86. https://doi.org/10.1016/j.nic.2014.03.003.
76. Martin RJ, Holthouse DJ, Wayne TG. Localising the source of pneumocephalus: a diagnostic problem. J Clin Neurosci. 2002;9(2) https://doi.org/10.1054/jocn.2001.0979.
77. Kurihara YY, Fujikawa A, Tachizawa N, Takaya M, Ikeda H, Starkey J. Temporal bone trauma: typical CT and MRI appearances and important points for evaluation. Radiographics. 2020;40(4) https://doi.org/10.1148/rg.2020190023.
78. Kwong Y, Yu D, Shah J, Kwong Y, Yu D, Shah J. Fracture mimics on temporal bone CT: a guide for the radiologist. Am J Roentgenol. 2012;199(2) https://doi.org/10.2214/AJR.11.8012.
79. Keshavamurthy VB, Ajith KM, Maradi N, Gupta R, Jain S, Keshavamurthy VB, et al. Correlation of hearing outcome in otic capsule sparing temporal bone fractures using temporal bone sub-site classification: a cross-sectional descriptive study. Egypt J Otolaryngol. 2022;38:1. https://doi.org/10.1186/s43163-022-00326-7.
80. Diaz RC, Cervenka B, Brodie HA. Treatment of temporal bone fractures. Journal of neurological surgery part B. Skull Base. 2016;77(5) https://doi.org/10.1055/s-0036-1584197.
81. Joyce T, Gossman W, Huecker MR. Pediatric abusive Head trauma. Treasure Island: StatPearls Publishing; 2023.
82. Liu Y-C, Chen I-C, Yin H-L, Wu Y-H, Lo S-H, Liang W-C, et al. Comparisons of characteristics and outcome between abusive head trauma and non-abusive head trauma in a pediatric intensive care unit. J Formos Med Assoc. 2023;122(11) https://doi.org/10.1016/j.jfma.2023.05.019.
83. Wright JN, Wright JN. CNS injuries in abusive Head trauma. Am J Roentgenol. 2017;208(5) https://doi.org/10.2214/AJR.16.17602.
84. Hymel KP, Wang M, Chinchilli VM, Karst WA, Willson DF, Dias MS, et al. Estimating the probability of abusive head trauma after abuse evaluation. Child Abuse Negl. 2019;88:266–74. https://doi.org/10.1016/j.chiabu.2018.11.015.

85. Goel G, Crawford J, Calayag M, Meltzer HS, Gonda DD, Levy ML. Mixed density subdural hematoma may not be pathognomonic for non-accidental trauma in infants. In: Neurosurgery USDDo, editor. Congress of neurological surgeons. San Diego, CA2016.
86. Wittschieber D, Karger B, Pfeiffer H, Hahnemann M. Understanding subdural collections in pediatric abusive Head trauma. AJNR. Am J Neuroradiol. 2019;40(3) https://doi.org/10.3174/ajnr.A5855.
87. Snelling PJ, Thanasingam AA, Jones P, Connors J. Comparison of abusive head trauma versus non-inflicted subdural haematoma in infants: a retrospective cohort study. Emerg Med Australas. 2022;34(6) https://doi.org/10.1111/1742-6723.14028.
88. Mankad K, Chhabda S, Lim W, Oztekin O, Reddy N, Chong WK, et al. The neuroimaging mimics of abusive head trauma. Eur J Paediatr Neurol. 2019;23(1) https://doi.org/10.1016/j.ejpn.2018.11.006.
89. Kriss S, Morris J, Martich V, Kriss S, Morris J, Martich V. Pediatric skull fractures contacting sutures: relevance in abusive Head trauma. Am J Roentgenol. 2021;217(1) https://doi.org/10.2214/AJR.20.23081.
90. Li X, Sandler H, Kleiven S. Infant skull fractures: accident or abuse?: evidences from biomechanical analysis using finite element head models. Forensic Sci Int. 2019:294. https://doi.org/10.1016/j.forsciint.2018.11.008.
91. Paiva WS, Oliveira AM, Andrade AF, Amorim RL, Lourenço LJ, Teixeira MJ. Spinal cord injury and its association with blunt head trauma. International journal of. General Medicine. 2011:4. https://doi.org/10.2147/IJGM.S15811.
92. Sheppard R, Kennedy GEM, Nelson A, Meguid EA, Darwish N. Vertebral artery injury in cervical spine fractures: a cohort study and review of the literature. Ulster Med J. 2020;89(2):89–94.
93. Martinez-Romo MA, McCoy CE. Jefferson fracture and the classification system for atlas fractures, a case report. J Educ Teach Emergen Med. 2021;6(2) https://doi.org/10.21980/J88P9C.
94. Turtle J, Kantor A, Spina NT, France JC, Lawrence BD. Hangman's fracture. Clinical Spine Surgery. 2020;33(9) https://doi.org/10.1097/BSD.0000000000001093.
95. Li J, Ho CL. Clinics in diagnostic imaging (214). Singapore Med J. 2022;63(9) https://doi.org/10.4103/SINGAPOREMEDJ.SMJ-2021-306.
96. Robinson A-L, Möller A, Robinson Y, Olerud C. C2 fracture subtypes, incidence, and treatment allocation change with age: a retrospective cohort study of 233 consecutive cases. Biomed Res Int. 2017; https://doi.org/10.1155/2017/8321680.
97. Gale SC, Gracias VH, Reilly PM, Schwab CW. The inefficiency of plain radiography to evaluate the cervical spine after blunt trauma. J Trauma Acute Care Surg. 2005;59(5) https://doi.org/10.1097/01.ta.0000188632.79060.ba.
98. Deluca A, Wichlas F, Deininger C, Traweger A, Mueller EJ, Deluca A, et al. Reevaluation of a classification system: stable and unstable odontoid fractures in geriatric patients – a radiological outcome measurement. Eur J Trauma Emerg Surg. 2022;48:4. https://doi.org/10.1007/s00068-022-01985-0.
99. Karamian BA, Liu N, Ajiboye RM, Cheng I, Hu SS, Wood KB. Reliability of radiological measurements of type 2 odontoid fracture. Spine J. 2019;19(8) https://doi.org/10.1016/j.spinee.2019.04.020.
100. Cho S, Shlobin NA, Dahdaleh NS. Os odontoideum: a comprehensive review. J Craniovertebr Junction Spine. 2022;13(3) https://doi.org/10.4103/jcvjs.jcvjs_71_22.
101. Moghamis IS, Asad W, Taha T, Head MO, Alhammoud A. Thoracic spine teardrop fracture – a case report. J Musculoskelet Surg Res. 2022;6(1) https://doi.org/10.25259/JMSR_63_2021.
102. Hall GC, Kinsman MJ, Nazar RG, Hruska RT, Mansfield KJ, Boakye M, et al. Atlanto-occipital dislocation. World J Orthop. 2015;6(2):236–43. https://doi.org/10.5312/wjo.v6.i2.236.
103. Utheim NC, Helseth E, Stroem M, Rydning P, Mejlænder-Evjensvold M, Glott T, et al. Epidemiology of traumatic cervical spinal fractures in a general Norwegian population. Injury. Epidemiology. 2022;9(1) https://doi.org/10.1186/s40621-022-00374-w.

104. Savakus JC, Weinberg DS, Moore TA, Vallier HA. Prevertebral soft-tissue swelling at C7 is highly sensitive for cervical spine ligamentous injury study type: retrospective cohort study. JAAOS Global Res Rev. 2020;4(4) https://doi.org/10.5435/JAAOSGlobal-D-19-00093.
105. Provenzale JM, Sarikaya B, Hacein-Bey L, Wintermark M, Provenzale JM, Sarikaya B, et al. Causes of misinterpretation of cross-sectional imaging studies for dissection of the Craniocervical arteries. Am J Roentgenol. 2012;196(1) https://doi.org/10.2214/AJR.10.5384.
106. Cox MW, Whittaker DR, Martinez C, Fox CJ, Feuerstein IM, Gillespie DL. Traumatic pseudoaneurysms of the head and neck: early endovascular intervention. J Vasc Surg. 2007;46(6) https://doi.org/10.1016/j.jvs.2007.08.021.
107. Merrill S, Clifton W, Valero-Moreno F, Damon A, Rahmathulla G. Vertebral artery injury with coinciding unstable cervical spine trauma: mechanisms, evidence-based management, and treatment options. Cureus. 2020;12(3) https://doi.org/10.7759/cureus.7225.
108. Renard D, Wacongne A, Parvu T, Thouvenot E. Radiological spectrum of pseudoaneurysms after extracranial internal carotid artery dissection. Stroke Vasc Interv Neurol. 2022;2(5) https://doi.org/10.1161/SVIN.121.000360.
109. Cruciata G, Parikh R, Pradhan M, Shah J, Greif E, Stein EG. Internal carotid artery dissection and pseudoaneurysm formation with resultant ipsilateral hypoglossal nerve palsy. Radiol Case Rep. 2017;12(2) https://doi.org/10.1016/j.radcr.2017.01.016.
110. Arnold M, Bousser MG, Fahrni G, Fischer U, Georgiadis D, Gandjour J, et al. Vertebral artery dissection. Stroke. 2006;37:10. https://doi.org/10.1161/01.STR.0000240493.88473.39.
111. Hosoya T, Adachi M, Yamaguchi K, Haku T, Kayama T, Kato T. Clinical and neuroradiological features of intracranial vertebrobasilar artery dissection. Stroke. 1999;30(5) https://doi.org/10.1161/01.STR.30.5.1083.
112. Mv B, Marchis GMD, Sarikaya H, Stapf C, Buffon F, Fischer U, et al. Differences and similarities between spontaneous dissections of the internal carotid artery and the vertebral artery. Stroke. 2013;44(6) https://doi.org/10.1161/STROKEAHA.113.001057.
113. Rao AS, Makaroun MS, Marone LK, Cho JS, Rhee R, Chaer RA. Long-term outcomes of internal carotid artery dissection. J Vasc Surg. 2011;54(2) https://doi.org/10.1016/j.jvs.2011.02.059.
114. Yaghi S, Engelter S, Brutto VJD, Field TS, Jadhav AP, Kicielinski K, et al. Treatment and outcomes of cervical artery dissection in adults: a scientific statement from the American Heart Association. Stroke. 2024;55(3) https://doi.org/10.1161/STR.0000000000000457.
115. Moriarty HK, Cearbhaill RO, Moriarty PD, Stanley E, Lawler LP, Kavanagh EC. MR imaging of spinal haematoma: a pictorial review. Br J Radiol. 2018;92(1095) https://doi.org/10.1259/bjr.20180532.
116. Kamo M, Watanabe Y, Numaguchi Y, Saida Y. Spinal subdural hematoma mimicking epidural lipomatosis. Magn Reson Med Sci. 2012;11(3):197–9.

Infectious, Inflammatory and Autoimmune, and Toxic

3

Grayson W. Hooper, Shruti Kumari, and Troy Holsworth

Infectious, Inflammatory and Autoimmune, and Toxic

Abstract

This chapter begins with important CNS infections, which are broken down into viral, bacterial, and fungal. The most important causes are first discussed, including herpes simplex virus, Human herpes virus 6, and HIV encephalitis, with their findings on CT and MRI. The review of HIV also includes a discussion of progressive multifocal leukoencephalopathy. Review of additional Flaviviridae

G. W. Hooper (✉)
Department of Neuroradiology, UH/Case Western Reserve University School of Medicine, Cleveland, OH, USA

S. Kumari
Department of Radiology, Case Western/UH, Cleveland, OH, USA

T. Holsworth
WVSOM, Lewisburg, WV, USA

G. W. Hooper (ed.), *The Neuronomicon*,
https://doi.org/10.1007/978-3-032-05385-5_3

infections like Japanese and tick-borne encephalitis follows due to their importance in Asia and Europe, respectively. The chapter then leans into MRI for distinguishing cerebritis from encephalitis, diagnosing subdural empyema and intraparenchymal abscess, recognizing ventriculitis, and finding evidence of meningitis. This section concludes with examples of unusual/opportunistic infections like tuberculosis meningitis and a brief discussion of prions.

The second section reviews demyelinating and inflammatory CNS diseases. A thorough discussion of MS begins the section, including the 2024 McDonald Criteria. Imaging distinctions between MS, NMO, and MOGAD are explored with MRI examples provided. Additional common demyelinating disorders like Guillain-Barré syndrome and ADEM are reviewed. This section is concluded with information about CNS vasculitis, neurosarcoidosis, and neuro-Behcet's disease.

The chapter concludes with toxic encephalopathy findings on MRI, including ethanol, methanol, ethylene glycol, opioid, and carbon monoxide poisoning, methotrexate toxicity, and hepatic encephalopathy.

There are forces at play in this chapter that require a plethora of Ph.D.s to understand, and even then, the body of knowledge remains murky to most. Aside from perhaps quantum mechanics, the immune system is one of *the most* complex topics in any field of science. Luckily, we as simple and humble neuroradiologists do not need to know more than a smidge of the source material. But what we must know cold is:

1. The anatomy
2. What normal looks like on CT and MRI
3. Pattern recognition of disease processes

Another important bit of information to know is clinical history. This includes acuity of presentation, symptoms, and *medication history*. I would have made this point #4 but I didn't want to screw up such a pristine number. As I've said multiple times before, we exist to add value to clinicians' searches for diagnoses and interventional planning. In the setting of a possible infection/inflammatory/autoimmune diagnosis, it's more of a two-way street. For example, FLAIR signal hyperintensity in the hippocampus and insula in a patient with fever and headache is vastly different than if it were discovered incidentally and without symptoms.

Joe Here Again

Imperative 1 That which restricts diffusion is not always infarct::mysterious glance::.

Imperative 2 Do not get tunnel vision based on a provided clinical history. Many of these disease processes have overlapping imaging features with similar diseases and even stroke. Closing the door on potential diagnoses with definitive language in your final report can be dangerous.

Imperative 3 Your reports should be more verbose than usual when describing these disease processes because the devil is in the details. For example, white matter lesions should be described by their exact location: juxtacortical, periventricular, etc. Cord signal abnormalities should be described by size, length, morphology, and position. These small details can significantly change the differential.

3.1 Basics

The discussion here will center on disease processes of exceptional lethality, morbidity, and medicolegal treachery.

- Viral Infection
  - Common Viral Encephalitides
    - Herpes Simplex Virus (HSV)
    - Human Herpes Virus 6 (HHV-6)
    - Tick-Borne Encephalitis
    - Japanese Encephalitis
    - HIV Encephalitis
      - Progressive Multifocal Leukoencephalopathy (PML)
  - Less Common Viral Encephalitides
    - Rabies Encephalitis
    - Measles Encephalitis: Subacute Sclerosing Panencephalitis (SSPE)
  - Varicella Zoster Ophthalmicus
- Bacterial Infection
  - Bacterial Meningoencephalitis
  - Cerebral Abscess
  - Subdural Empyema
  - Ventriculitis
  - Mycotic Aneurysm
  - Peritonsillar Abscess
  - Ludwig's Angina
- Fungal/Atypical Infection
  - Tuberculous Meningoencephalitis
  - Cryptococcal Meningoencephalitis
  - Toxoplasmosis
  - Rhinocerebral Mucormycosis
  - Prion Diseases
- Demyelinating/Inflammatory/Autoimmune
  - Demyelinating Diseases
    - Multiple Sclerosis (MS)
    - Acute Disseminated Encephalomyelitis (ADEM)

    - Acute Hemorrhagic Leukoencephalitis (AHL)
  - Guillain-Barré syndrome (GBS)
  - Myelin Oligodendrocyte Glycoprotein Antibody Disease (MOGAD)
  - Neuromyelitis Optica Spectrum Disorder (NMO-SD)
  - TNF-Alpha Blocker-Associated Demyelinating Disorder
- Inflammatory and Autoimmune Disease
  - Susac's Syndrome
  - Neurosarcoidosis
  - Neuro-Behçet's Disease (NBD)
  - Primary Angiitis of the CNS (PACNS)
  - Immune Reconstitution Inflammatory Syndrome (IRIS)
- Toxic Encephalopathy
  - Carbon Monoxide Poisoning
  - Ethylene Glycol Poisoning
  - Methanol Poisoning
  - Methotrexate
  - Radiation Induced
  - Opioids
  - Ethanol
  - Hepatic Encephalopathy

3.2 Viral Infection

It is both a philosophical and biological open debate as to whether viruses can even be considered living entities. Regardless, these little packages of nucleic acid information are capable of wreaking extraordinary havoc on the human central nervous system.

There is a huge number of viruses out there capable of infecting human beings, and that number is always changing. Viruses frequently mutate and new ones emerge. Those with CNS tropism have variable clinical presentations ranging from a simple headache to coma and death. In this section, we'll discuss some more specific imaging appearances. For example, Japanese encephalitis and HSV are common causes of encephalitis globally and in Western countries, respectively, and each has some suggestive imaging findings [1]. However, understand that many are completely nonspecific. Therefore, taking a holistic approach to diagnosis will save you immense frustration from not being able to suggest a specific pathogen. In many instances, the laboratory nerds will also not be able to identify the patient's infectious agent.

3.2.1 Bottom Line Up Front

Viral encephalitis may exist alone, but it more often presents with accompanying meningitis. The clinical manifestations are highly variable depending on the specific individual involved as well as the specific pathogen. Remember when I said immunology is insanely complex? This is just one example. A virus that irritates one person may kill another [2]. Except rabies. And herpes. They definitely just kill.

Five Diagnostic Pearls for the Resident/Fellow on Call:

- If you see isolated hypoattenuation in one or both medial temporal lobes on a CT examination, call the referring physician and proffer possible HSV encephalitis, especially if you are told they are acutely ill. Stroke is in the differential diagnosis, but MRI can come after they start treatment.
- When reviewing the brain MRI, ensure that you carefully scrutinize the sulci and gyri along the apex convexities. Subarachnoid FLAIR signal hyperintensity is what you seek.
- Be aware that a subarachnoid FLAIR signal has a differential diagnosis, which includes anomalous false positives caused by susceptibility artifact, slow flow in pial vessels, or when the patient is on supplemental oxygen.
- Use your gradient/SWI sequences! Microhemorrhages are associated with HSV encephalitis and other aggressive necrotizing processes.
- There is a differential diagnosis for abnormal signals in the temporal lobes/hippocampi.

1. Common Viral Encephalitides

   The top three viral encephalitides are HSV, tick-borne encephalitis (TBE), and Japanese encephalitis. We have had a brief introduction to the first and the third, but what about TBE? Well, TBE is vectored by the vile aberration known as the tick, which is a parasitic arthropod, but it can also be acquired by consuming raw milk. Ticks are responsible for vectoring a multitude of illnesses beyond the scope of this book; however, for our discussion, they vector Flaviviridae. TBE is a significant emerging health threat found in Europe [3].

   - **HSV Encephalitis** is a critical, "no miss" diagnosis, as it is the most common encephalitis in Western countries, and early initiation of antiviral treatment is critical to mitigate morbidity and mortality [1, 4].
     - HSV encephalitis can demonstrate low attenuation in the temporal lobes and insular cortex, but CT findings can also be completely absent [5].
     - MR imaging is essential in the diagnosis of HSV encephalitis. Common imaging findings include:
       - FLAIR/T2 hyperintensity in the bilateral temporal lobes in the proper clinical context is nearly pathognomonic [6]. The cingulate cortices are often involved as well.
       - Postcontrast enhancement is common in HSV encephalitis.
       - HSV encephalitis often spares the basal ganglia [7].
       - Restricted diffusion may be present, and bilateral parenchymal restricted diffusion is associated with worse outcomes, as is restricted diffusion that extends into the thalamus [8].

    - **Microhemorrhages observed on T2* and SWI are suggestive of HSV encephalitis. Microhemorrhages effectively rule out postictal edema.**
  - There are many MRI mimics of HSV temporal lobe encephalitis (Fig. 3.1), and differentiating them can be challenging. Some of these mimics are also infectious, whereas others are not [9] (Figs. 3.2 and 3.3).
    - Tuberculosis and varicella-zoster virus are common infectious agents that can mimic the disease.
    - Noninfectious causes include ADEM, gliomas, NMDA receptor limbic encephalitis, CNS vasculitis, and paraneoplastic syndromes (Fig. 3.2).
- **HHV-6** may trigger memories of roseola infantum from your pediatrics rotation, a relatively benign condition. However, the causative virus, which infects a majority of the world's population, tends to chill out after the initial infection. It can later recur as encephalitis in both immunocompetent and immunocompromised individuals, and it can be nasty, though it most often presents in patients who have undergone hematopoietic stem cell transplant. Therefore, a specific diagnosis is essential because HHV-6 does not respond

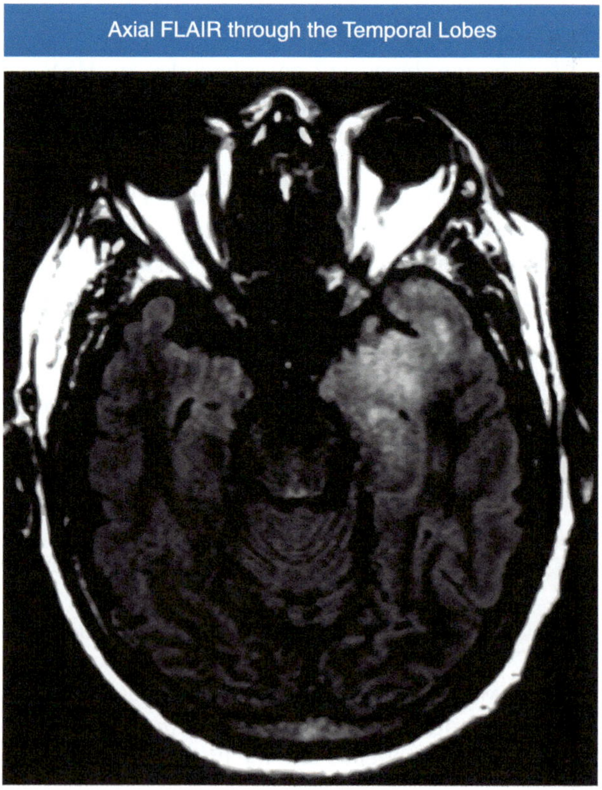

Fig. 3.1 An axial FLAIR MR image through the temporal lobes reveals significant left temporal lobe FLAIR signal hyperintensity and swelling. Faint signal hyperintensity is present in the right temporal lobe. You should notify the ordering provider immediately of these findings

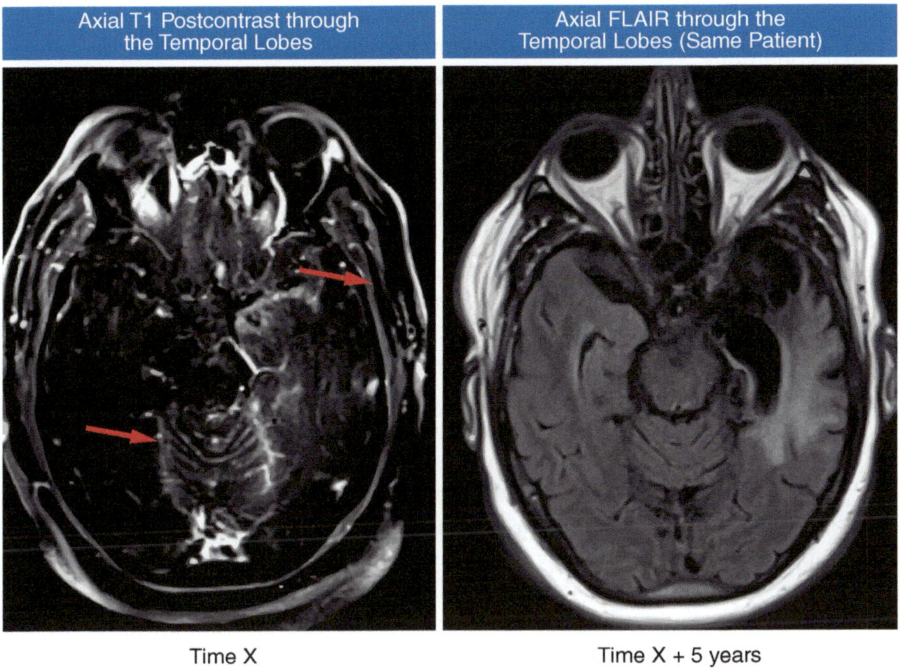

Fig. 3.2 Annotated axial T1 postcontrast imaging on your left reveals avid cerebellar folia and left temporal lobe meningeal enhancement. This patient was acutely ill with significant neurologic deficits. Repeat brain imaging 5 years later is seen on your right. Notice the left temporal lobe encephalomalacia and gliosis. This lucky patient thankfully survived acute herpes meningoencephalitis

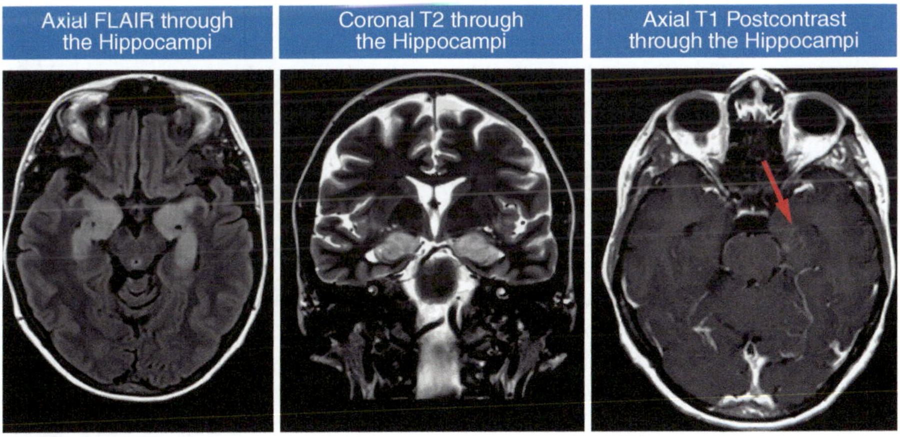

Fig. 3.3 MR imaging through the brain of a patient with memory deficits. The patient's presentation was non-acute and afebrile. Though the imaging findings, which include hippocampal FLAIR and T2 signal hyperintensity as well as postcontrast enhancement, are concerning, this was a case of limbic encephalitis. Many cases of limbic encephalitis are secondary to paraneoplastic effects from an underlying malignancy, but this case was spontaneous

well to acyclovir, which is normally used to treat HSV encephalitis, **but rather HHV-6 responds to ganciclovir** [10].

- Differentiating HHV-6 from HSV encephalitis can be difficult in the early phases of the disease, and lab testing is usually required for definitive diagnosis. There are overlapping features between the two on CT and MRI. However, the *absence* of others is quite useful [11, 12].
  - CT findings can initially be absent.
  - MRI findings can initially be absent.
  - FLAIR/T2 signal hyperintensity will be exclusively seen in the mesial temporal lobes with HHV-6 infection.
  - If there is *extra-temporal* involvement, HHV-6 is unlikely.
  - CT findings are more likely to develop later in the disease course of HSV encephalitis.

- **TBE** is a growing public health concern in Europe but rarely seen in the US, aside from travelers and military personnel who have been in endemic areas [13]. After initial infection, 70% of patients will remain asymptomatic but the remainder will experience a nonspecific viral prodrome. Of those patients, approximately one-third will develop neurologic sequelae, including meningitis, encephalitis, and/or radiculitis, which can be very severe [14].
  - The imaging findings of this disease process are variable, both on CT and MRI, but some lesions can hint at the diagnosis [15, 16].
    - CT imaging can demonstrate loss of gray-white differentiation in the basal ganglia. Of course, the differential is MCA infarct.
    - MRI findings (FLAIR signal hyperintensity) in the brain tilt toward the cerebellum, thalami, basal ganglia, and midbrain.
      - Pulvinar involvement of the thalamus can look like Creutzfeldt-Jakob disease (CJD).
      - Involvement of the basal ganglia can mimic toxic metabolic disorders.
      - This disease has a predilection for the ventral horn gray matter in the spinal cord, which can result in an "owl's eyes" appearance on axial FLAIR.
- **Japanese encephalitis**, unlike the Flaviviridae vectored by ticks, gets its ride from *Culex* mosquitoes. It is endemic throughout Asia and Northern Australia, even though only 1% of those infected will develop encephalitis [17]. There is an underutilized vaccine (Ixiaro) available for this disease.
  - MRI and CT characteristics for Japanese encephalitis have been described and are similar to other Flaviviridae [18, 19].
    - Similar to TBE and HSV encephalitis, CT findings are less frequently observed than MRI findings.

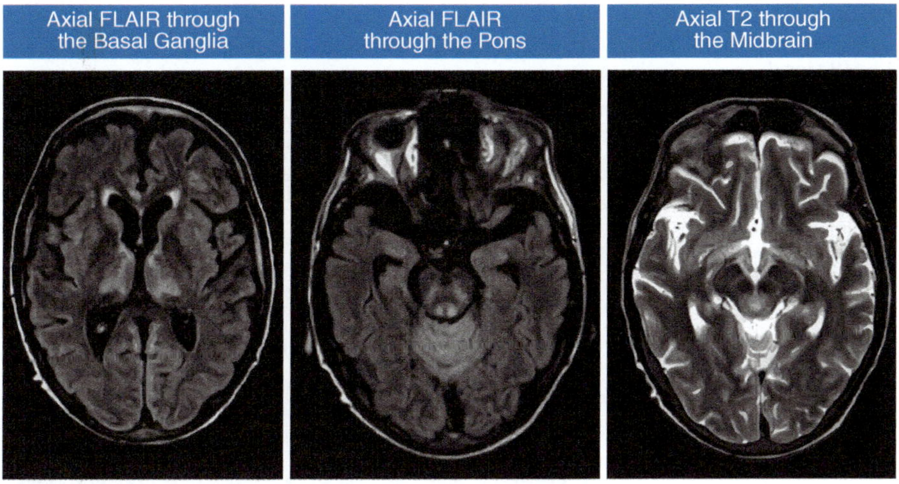

Fig. 3.4 This patient initially presented with a headache accompanying an acute systemic illness. They rapidly became obtunded. MR imaging reveals FLAIR signal hyperintensities within the bilateral thalami, cerebellum, and periaqueductal gray matter. These findings were suspicious for Japanese Encephalitis or other Flaviviridae infection, but a definitive diagnosis was never established

- Thalamic T2/FLAIR signal hyperintensity seems to be the imaging hallmark of Japanese encephalitis (Fig. 3.4).
  - It is also commonly found in dengue encephalitis.
  - Many other viruses, including arboviruses, also have similar imaging appearances.
- Japanese encephalitis can present atypically with medial temporal lobe involvement, which can be difficult to differentiate from HSV.

- **HIV encephalitis** is historically not a straightforward diagnosis. The virus can itself directly injure neurons, resulting in neurocognitive decline, or it can make the patient susceptible to opportunistic infections. But HIV can also cause a number of inflammatory disorders. For a lengthy (but interesting!) discussion on its history, I recommend the 2022 Journal of Neurology article by Dr. Lucas [20].
  - If you are reviewing the MRI of a patient with neurologic deficits, ask to know the CD4+ T-cell count. If you hear a number below 200, increase your index of suspicion and broaden your differential.
  - The imaging features of HIV encephalitis on MRI [21, 22]:
    - Diffuse symmetric periventricular white matter FLAIR signal hyperintensities that spare the cortical U-fibers.
    - **Cortical atrophy is the most common finding, and it is out of proportion to the patient's age**.
    - There is no contrast enhancement (Figs. 3.5 and 3.6).

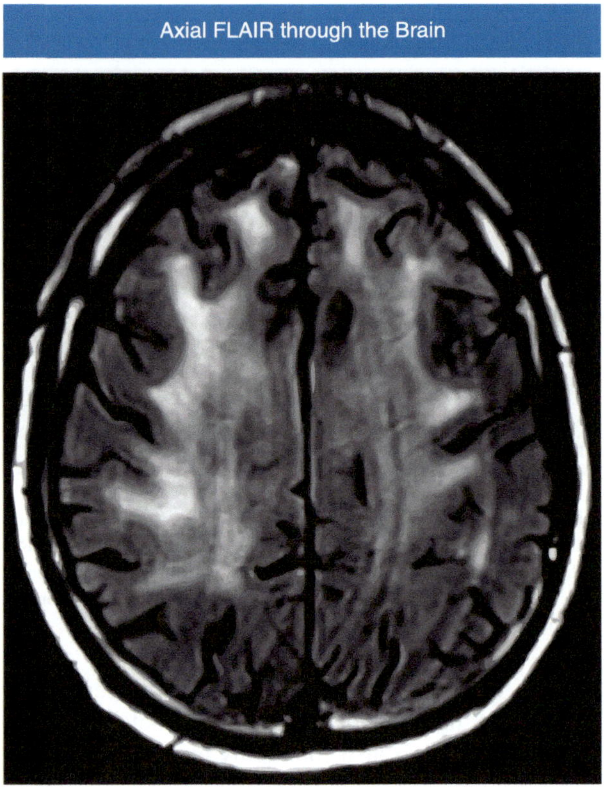

Fig. 3.5 This axial MRI image reveals patchy FLAIR signal hyperintensities bilaterally that appear to spare the cortical U-fibers. The patient was in their early 30s; thus, the sulcal and gyral patterns suggest atrophy out of proportion to the patient's age. These findings are common in HIV encephalitis

- **Progressive multifocal leukoencephalopathy** is a vicious disease that occurs in immunocompromised patients. In the past, it was mostly HIV patients with low T-cell counts (50–100 cells/microliter) who suffered; however, it's now more frequently seen in patients after stem cell transplants. **PML can also be seen in patients with MS who are treated with natalizumab**. It is caused by reactivation of the JC (John Cunningham) virus, which chills out in immunocompetent hosts, but ravages the immunocompromised. Without treatment, it is universally fatal. Even with treatment, the mortality is high, and the morbidity for the survivors is higher.
  - MRI can suggest PML, as there are some characteristic findings [23, 24]:
    - Demyelinating lesions of different ages show different features
      - (i) Older lesions are more uniformly T2 hyperintense.
      - (ii) Newer lesions are punctate, irregular, and T2 hyperintense.
        - (1) Galaxy/Milky Way sign
    - Occipital and parietal lobes are often involved, and so too are the subcortical U-fibers.
    - T1 isointensity or hypointensity. Enhancement is possible but unusual in PML, unless the patient has some functional immune system.

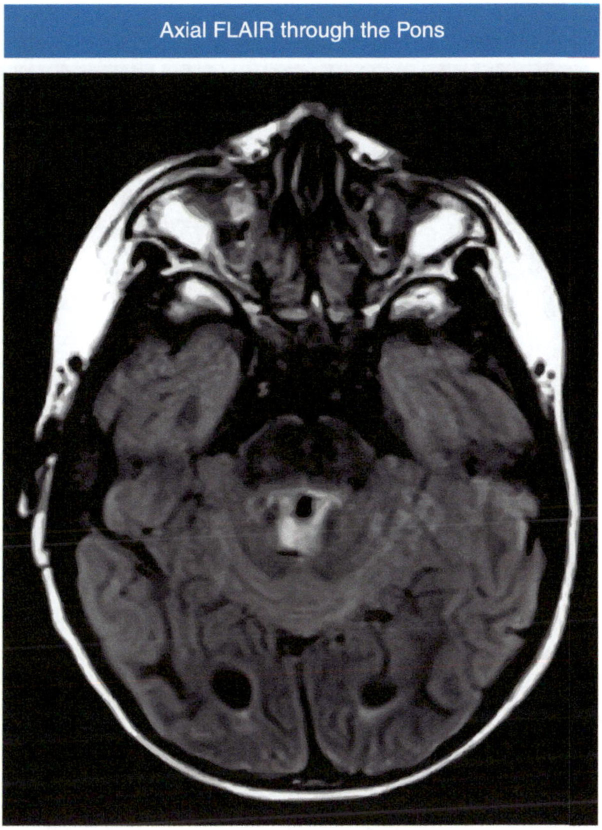

Fig. 3.6 An axial FLAIR image through the pediatric pons demonstrates FLAIR signal hyperintensities in the cerebellum and dorsal pons (hindbrain). Rhombencephalitis is the diagnosis in this pediatric patient. The most common causes are HSV, *Listeria* species, and adenovirus. As far as encephalitides go, this one has a high mortality rate

(i) Immune reconstitution (discussed later) while in an active PML phase can produce striking worsening of imaging findings.

– Leukocortical "band" of SWI hypointensity can be seen at the periphery of these lesions. Look for wavy gray-white junction black lines (Fig. 3.7).

2. Less Common Viral Encephalitides

Welcome to the world of the weird. There are a few special viral pathogens that we'll discuss here. Though infectious disease is a fascinating topic, and there are hundreds upon hundreds of bad ways to die, we need to remain focused.

- **Rabies encephalitis** means death. There's just no other way to state it. Reports of survival emerge every *decade* or so but for all intents and purposes, patients who show signs of rabies are doomed. Though quite rare in the United States, rabies causes around 60,000 deaths per year around the world, most commonly in developing nations, and it is frequently vectored by dogs [25, 26].

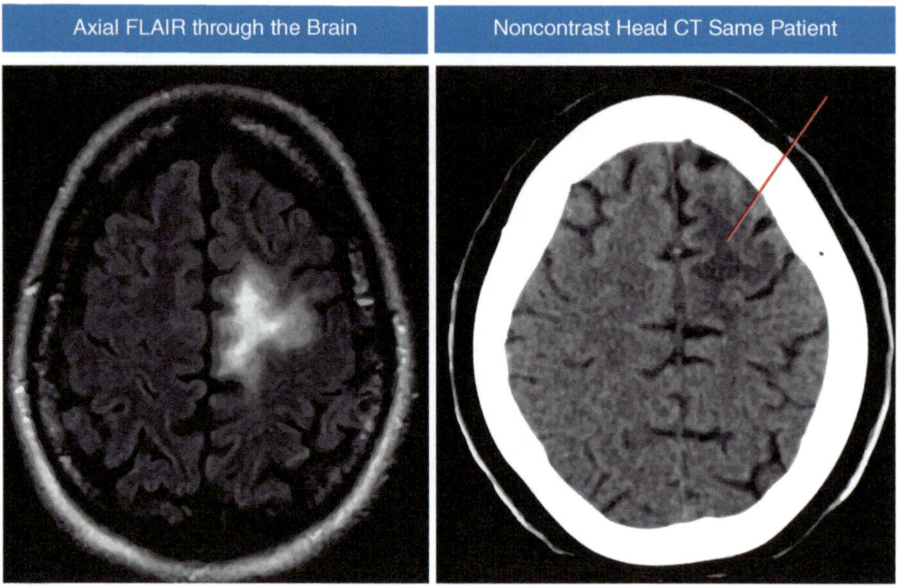

Fig. 3.7 Axial MRI and annotated CT imaging through the subcortical white matter of the frontal lobes reveals FLAIR signal hyperintensity (MRI image: left) and CT hypodensity (CT image: right) that involves the cortical U-fibers. In an immunocompromised patient, PML is a serious concern. ADEM and PRES could also be considered

- MRI is occasionally obtained in patients in the fulminant course of the disease [27, 28]:
  - FLAIR/T2 signal abnormalities localize to the brainstem, thalami, and basal ganglia (similar to Japanese encephalitis).
  - The paralytic form of the disease can cause swelling of the cord and FLAIR signal hyperintensity, **which can mimic GBS**.
  - Remember—this is a disease of gray matter.
  - Any animal bite needs postexposure prophylaxis (PEP). Hell, even waking up in a room with a bat should get the patient PEP treatment, regardless of whether bites are present or not. Yes, I know this statement is controversial, but if it were me or my loved ones, I'd opt for treatment 100% of the time.

- **Measles encephalitis** is kind of a rare bird that occurs after rubeola (measles) infection, which should by all common decency and logic *no longer exist* but somehow does. It is subacute in presentation, but the term subacute in this setting is something of a misnomer, as symptom onset is usually 6 years after measles exposure [29]. While equally lethal to rabies, SSPE is slow, insidious, and rare, affecting 1 in 100,000 patients infected with measles.

- MR imaging findings [30, 31]:
  - Parieto-occipital white matter FLAIR signal hyperintensities are common. Kind of like PML, no?
  - Gray matter involvement is rare.
  - Does not enhance.

3. Varicella Zoster Ophthalmicus (VZO)

   VZO, also known as herpes zoster ophthalmicus, is a disease process that occurs after latent reactivation of the chickenpox virus. If you recall your anatomy and pathology, astutely, you'll realize that herpes viruses enjoy setting up shop in the trigeminal ganglia. Occasionally, they'll snowbird in V1, V2, or V3, creating a nasty vesicular rash in those distributions. Unfortunately, the virus can spread, causing uveitis, scleritis, conjunctivitis, or optic neuritis in rare instances [32]. VZO is an emergency. It can cause permanent vision loss.

   - MRI findings can vary depending on the structures involved in the infection [33, 34].
     - Given that this is a trigeminal-based disease, look for enhancement of the intracisternal nerve segments, the Gasserian ganglion, and CNV_1.
     - The subcutaneous soft tissues can be edematous and enhancing.
     - The lacrimal gland can swell if dacryoadenitis ensues.
     - Rarely, an abnormal signal can be seen in the medulla—the so-called *trigeminal tract sign*.
   - Varicella Zoster encephalitis also exists. It is an unusual virus and has some bizarre manifestations. For example, it is most commonly seen in immunocompromised patients and can have a tropism for the cerebellum. However, it is primarily a vasculopathy of large and small vessels, which can ultimately result in stroke [35, 36].
     - MRI findings are variable and dependent on whether it's the direct viral tropism of CNS cells or a resulting vascular insult.
       - Common imaging findings include subcortical clustered FLAIR signal hyperintense plaques that can enhance.
       - If you see discrete restricted diffusion in the basal ganglia in a young patient with concomitant vascular insult on MRA, think varicella zoster, which **is a significant cause of pediatric stroke** [37].
       - On at least one occasion, the disease struck a patient's corpus callosum, and the MRI findings mimicked glioblastoma.
   - When varicella zoster reactivates in the facial nerve, specifically the geniculate ganglion, look for the triad of facial paralysis, otalgia, and a vesicular rash in the external auditory canal [38]. **This disease process is called Ramsay-Hunt syndrome, and it is a favorite pimp question from wily attendings**.

3.3 Bacterial Infection

There is no debate here—bacteria are living creatures, and many of them suck. If one is immunocompromised, a whole bunch more of them suck. How exactly they infect the CNS is fairly straightforward: hematogenous or direct extension. We'll discuss some examples of each. Moreover, we'll discuss the various manifestations of infection and their imaging characteristics.

3.3.1 Bottom Line Up Front

Though viral encephalitis is most common, bacterial encephalitis does exist. *Mycobacterium tuberculosis* and *Listeria* species are implicated—refer to rhombencephalitis above in Fig. 3.6. Clinical distinction is made between encephalitis and meningoencephalitis, though differentiating them is quite difficult. Enteroviruses are actually the most common causes of viral meningitis and usually have a good outcome. However, this point is important: bacterial meningitis has a high mortality, whereas viral meningitis is often self-limiting, with the exception of HSV [39]. Not to say that other encephalitides are a walk in the park, but bacterial meningitis is far worse. With respect to imaging, we're going to discuss their differences.

Five Diagnostic Pearls for the Resident/Fellow on Call:

- Some folks erroneously use the terms cerebritis and encephalitis interchangeably. This is not accurate. Cerebritis is more often a **focal** infection of the brain parenchyma, often bacterial. Encephalitis is more commonly a **generalized** inflammation of the brain, often viral. Be discerning in your language.
- Infections, including encephalitis and cerebritis, can mimic stroke in the acute setting on CT.
- To further muddy the waters, CNS infections can precipitate stroke.
- Always check the paranasal sinuses and the teeth on your initial head CTs.
- Always check the mastoid air cells on your initial head CTs.

1. Bacterial Meningoencephalitis

   A young student arrives by EMS with fever, a stiff neck, and altered mental status. Though only 40% or so of patients present in this manner, it is nonetheless regarded as "textbook," and the information you've gleaned from the ER attending is useful [40]. The two main offenders are Neisseria *meningitidis* and Streptococcus *pneumoniae*. In young and unvaccinated individuals, *Haemophilus influenzae* is possible. For patients from developing nations or prisons, consider *Mycobacterium tuberculosis*.

   - First and foremost, **lumbar puncture (LP) should be prioritized**. The care team, including you, should avoid delaying the LP in order to obtain neuroimaging, unless necessary. CT may be indicated before LP if there are clinical signs of intracranial hypertension, or if the patient has neurologic deficits or is obtunded. **In one study, only 8/549 (1.5%) CT scans had findings that affected clinical management** [41, 42].

- Consider initial neuroimaging in patients who are immunocompromised. Their pretest probabilities of having a contraindication to LP are higher.
- MRI of patients with meningitis can demonstrate subarachnoid FLAIR signal hyperintensity, and irregular and thick meningeal enhancement. It can also be useful for assessing meningitis mimics [43]. Remember, subarachnoid FLAIR signal hyperintensity has a differential.
- Even in the acute setting, CNS infection can mimic stroke on CT [44].
- Bacterial meningiti/meningoencephalitis can cause stroke, especially in the immediate vicinity of the infectious material [45] (Fig. 3.8).

? Cerebral Abscess

Obviously, not an ideal situation, a cerebral abscess is a quarantined immunohistochemical warzone where the battlefield is the brain parenchyma. It can be the natural evolution of direct extension of cerebritis/meningoencephalitis, or it can be the result of hematogenous deposition of bacteria in an otherwise healthy brain. The most common causative organisms of brain abscesses are *Streptococci* and *Staphylococci* species, and patients will commonly have predisposing conditions [46]. Otomastoiditis, dental abscess, right-to-left flow shunts, and endocarditis are all predisposing conditions.

- Imaging characteristics of cerebral abscess depend on which phase of the disease process imaging is obtained [47–49].
  - Initially, CT imaging will show decreased attenuation. It can mimic an infarct.

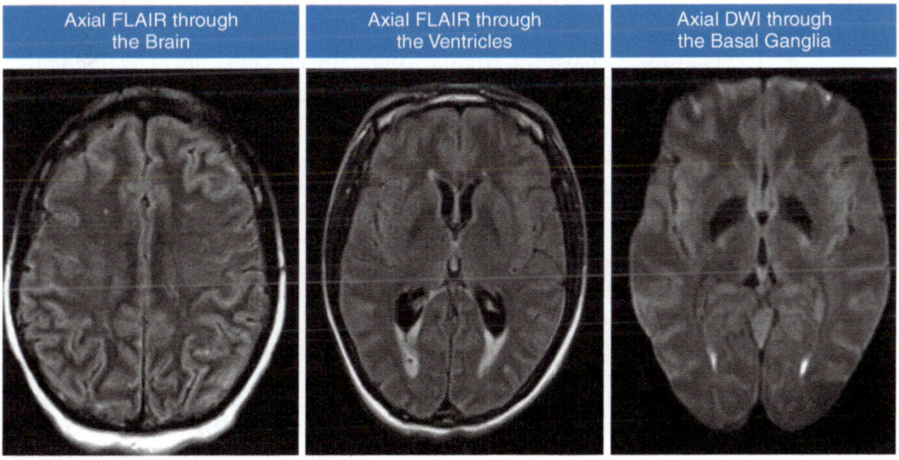

Fig. 3.8 Axial FLAIR images on your left and center reveal non-suppression of FLAIR signal within the cortical sulci and layering material in the atria of the lateral ventricles respectively. The right DWI image shows restricted diffusion in the dependent ventricles. These findings are strongly suggestive of meningoencephalitis with concomitant ventriculitis

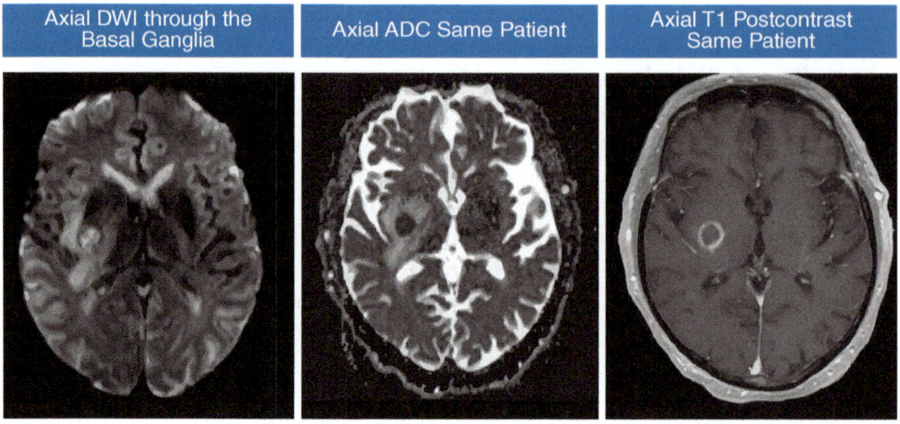

Fig. 3.9 Multisequence imaging through the basal ganglia reveals a ring-enhancing mass that restricts diffusion centrally. There is surrounding vasogenic edema but minimal mass effect. The patient was an intravenous drug abuser, a common cause of hematogenous dissemination of bacteria

- As time progresses, the infected parenchyma will liquefy (barf) and develop an enhancing ring.

– MRI has better sensitivity and specificity for the disease process, and it is here that restricted diffusion really is your friend.

- The liquefied parenchyma will demonstrate T2 signal iso- or hyperintensity. It can also show T1 signal hyperintensity.
- The liquefied core will restrict diffusion in most **central** instances.
- A thick enhancing ring represents the capsule. Note that abscess formation is more common in immunocompetent patients.
- Expect to see vasogenic edema around the abscess core. The mass effect can be significant, so assess for herniation, ventricular effacement, and midline shift.
- As always, nothing is of absolute certainty in this field. Necrotic tumors can have similar appearances to brain abscesses. Abscesses usually show **central restricted diffusion**. If the case remains unclear, perfusion MRI can be helpful in distinguishing the two.
- Tumefactive demyelination is lower on the differential. The ring enhancement in this disease process is often partial, and the restricted diffusion often occurs at the leading edge of demyelination (Fig. 3.9).

It is here that I wish to make an important point. A point so important that it gets its own line. **Subacute hematomas can restrict diffusion in a striking manner.** Always corroborate with your other sequences, namely, precontrast T1.

3. Subdural Empyema

When purulent material accumulates in the extra-axial space, this is what occurs. It is a serious complication of bacterial meningitis, especially in newborn patients, and may result in permanent neurological deficits if it doesn't kill the patient first. Therefore, it is considered a neurosurgical emergency [50, 51]. Whereas subdural empyema is most commonly caused by meningitis in newborns, it is most commonly the result of direct extension of disease from the sinuses/teeth or mastoids in adults [52].

- Imaging characteristics of subdural empyema may overlap with other subdural collections on CT and, to a lesser extent, on MRI [52–54].
  - CT imaging can show an extra-axial fluid collection of varying density that can look like a hematoma.
  - MRI has the added benefit of demonstrating enhancement and encapsulation of the extradural collection, as well as restricted diffusion. Note that atypical organisms like Escherichia coli can demonstrate a heterogeneous DWI signal that may confuse the picture.
- Subarachnoid empyema is also possible. We know this as bacterial meningitis, as discussed above. However, it is absolutely possible to detect subarachnoid restricted diffusion if pus accumulates in this space. If you see this, raise the possibility of purulent material, blood, or malignant cells (Figs. 3.10, 3.11, and 3.12)!

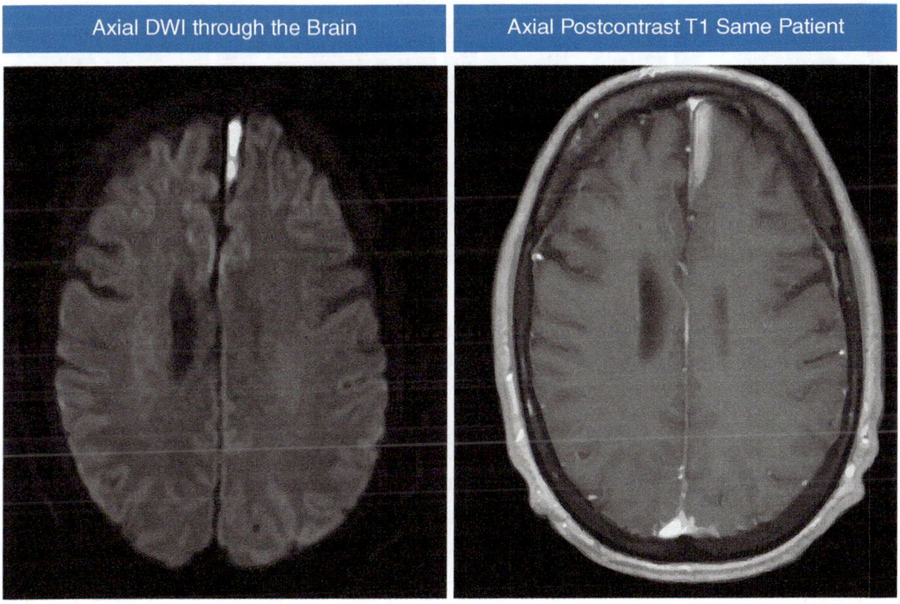

Fig. 3.10 On the left, we have DWI signal hyperintensity along the left parafalcine frontal lobe. Postcontrast T1 imaging reveals an oblong T1 signal hyperintense mass corresponding to the restricted diffusion. Diagnosis? Subacute subdural hematoma. Make sure to check your precontrast T1 imaging to confirm that the signal hyperintensity you see is enhancement ***or*** precontrast T1 intrinsic, which it was in this case (precontrast T1 not shown)

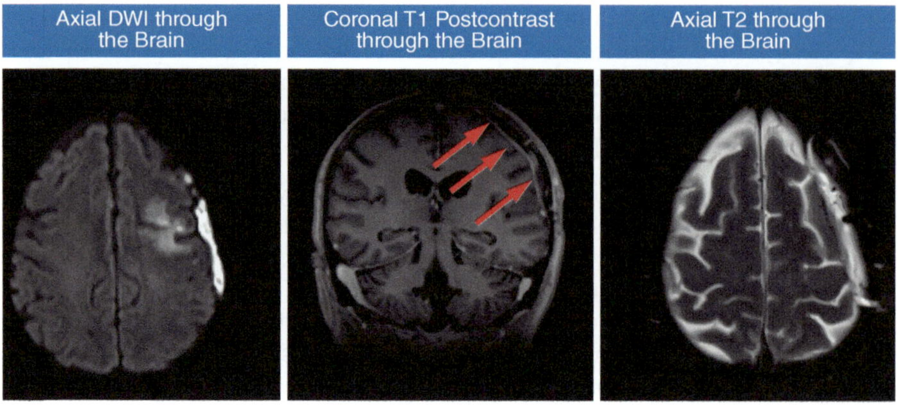

Fig. 3.11 In this recently postoperative patient, the far-left image reveals DWI signal hyperintensity in the left frontal extra-axial space. Postcontrast imaging shows thick peripheral enhancement. The far right axial T2 image shows a discrete fluid collection. These findings are most concerning for a subdural empyema

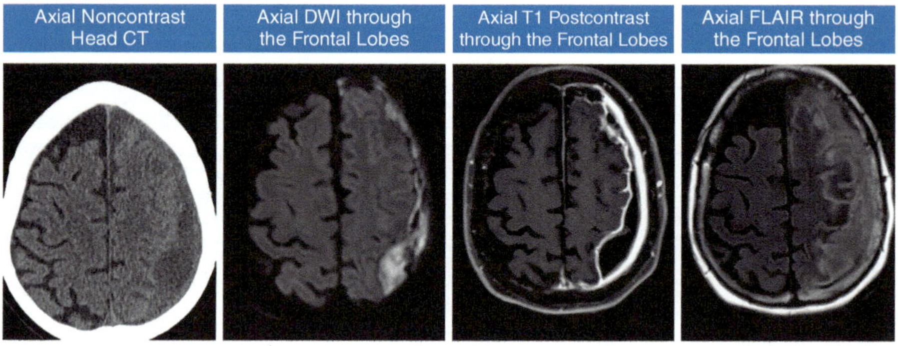

Fig. 3.12 A florid left subdural empyema with additional meningoencephalitis. The far-left axial CT image shows hazy material along the left extra-axial space. Much of this material restricts diffusion. The axial postcontrast image reveals thick peripheral enhancement. The far-right axial FLAIR demonstrates failure of signal suppression in the cortical sulci

4. Ventriculitis

   Inflammation of the ependymal lining is a severe consequence of CNS infection, which portends a poor prognosis. Despite a 30% mortality and 60% incidence of neurologic sequelae, ventriculitis does not have a common definition [55]. Furthermore, patients possessing an extraventricular drainage device (EVD) may show signs of ventriculitis that are inflammatory as opposed to infectious [56]. The most common causes of ventriculitis are meningitis and surgical intervention, but hematogenous seeding of the choroid plexus can also occur. If the patient is an adult and immunocompromised, think cytomegalovirus (CMV) ventriculitis.

- Imaging of ventriculitis has some distinct features, though it can overlap with others, especially on CT [43, 57].
  - CT imaging is nonspecific, but the presence of layering materials in the dependent ventricles can suggest pus (or blood).
  - Hydrocephalus.
  - MR imaging may demonstrate avid ependymal enhancement, layering material, and sometimes septations.
  - Restricted diffusion can be seen within the ventricles, but this may be variable due to purulent material mixing with normal CSF.
  - Ventriculitis is the most common CNS manifestation in CMV reactivation. Ependymal enhancement and diffusion restriction in this setting portend a very poor outcome (Fig. 3.13).

5. Mycotic Aneurysm

   The term "mycotic" in this disease process does not necessarily mean fungal. It rather refers to the shape of the aneurysm, as they look like mushrooms according to the person who coined the term. In reality, most mycotic aneurysms are bacterial in origin. They occur when bacteria escape the vasa vasorum into the adventitia and grow inwards, which weakens the vessel wall against normal

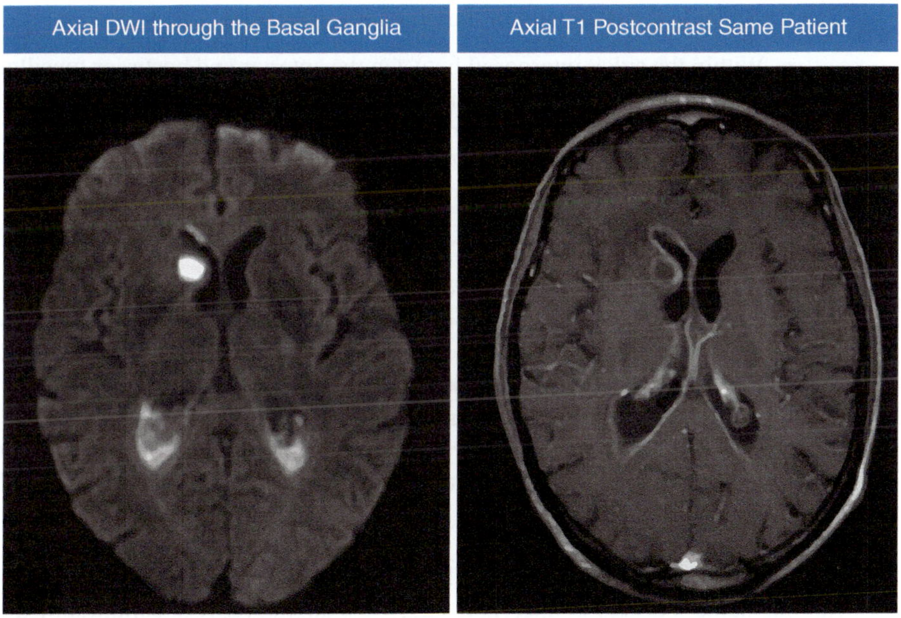

Fig. 3.13 A peripherally enhancing mass that restricts diffusion centrally is seen in the right caudate head. Additional diffusion-restricting material is seen in the atria of the lateral ventricles. The postcontrast axial T1 image on your right demonstrates subependymal enhancement along the right-greater-than-left lateral ventricles. Diagnosis? Intraparenchymal abscess and ventriculitis with purulent material within the ventricles

blood flow, allowing fusiform growth [58]. Infectious endocarditis is the most common cause of mycotic aneurysms.

- Both MRI and CT have moderate sensitivity in the diagnosis of mycotic aneurysms, though CT appears to shine in this setting [59–61].
  - Mycotic aneurysms may develop rapidly over time; therefore, short-interval aneurysm formation suggests this diagnosis.
  - Multiplicity of aneurysms occurs in up to 25% of cases.
  - These aneurysms tend to be more distal in the intracranial circulation.
  - **DSA/cerebral angiography is the gold standard for detecting mycotic aneurysms.**

6. Peritonsillar Abscess

These will roll across your list on a regular basis and are usually straightforward imaging diagnoses. The key to diagnosing peritonsillar abscesses (PTAs) is having a high-quality, IV contrast-enhanced CT through the neck. Occasionally, the scan will be marred by beam hardening and streak artifact in patients with dental amalgam or other metallic hardware.

- Head and neck anatomy can be a beast. Therefore, it is important that you learn the visceral spaces of the neck as well as the cervical lymph node stations.
- Localize a PTA by first finding the uvula on axial imaging.
  - You'll be in the proximity of the glossotonsillar sulci.
    - Within the sulci, interposed between the anterior and posterior tonsillar pillars, live the palatine tonsils.
    - Scrutinize these areas for peripherally enhancing fluid collections.
    - If there are no fluid collections, look for distortion of normal fat planes, such as the parapharyngeal space fat.
  - Assess for cervical station lymphadenopathy.
    - Big reactive lymph nodes are normal in this setting. Measure a few in short axis. Greater than 10 mm is generally considered pathologically enlarged (Fig. 3.14).
    - **If the patient is over age 30, you must be aware that p16+ (HPV-related) tonsillar squamous cell carcinoma may be lurking.**
      - **A cystic mass in a cervical lymph node station in any patient over age 30 must be eyed suspiciously. It could be a cystic nodal metastasis.**

7. Ludwig's Angina

As the story goes, old man Ludwig may have died of the very disease that bears his name, but this is uncertain. Ludwig's angina (LA) is cellulitis of the floor of the mouth that traverses multiple spaces to include **the sublingual space,**

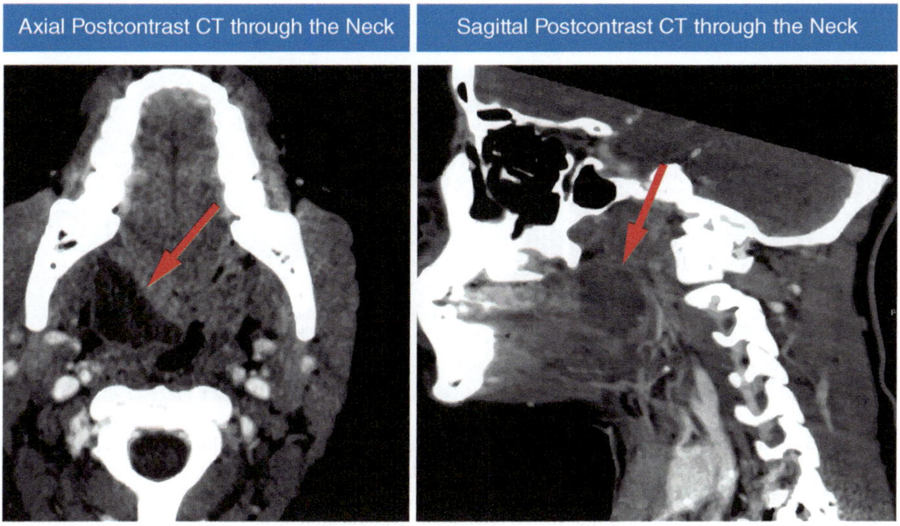

Fig. 3.14 Annotated axial and sagittal CT images reveal an organizing fluid collection in the right peritonsillar soft tissues. One would report the size of the collection and describe the obvious mass effect on the pharyngeal airway. For a complete report, you should include information regarding cervical station lymphadenopathy if present

submandibular space, and submental space. Make no mistake, this disease process can be challenging to diagnose, and mortality exceeds 50% if left to its own devices.

- Inflammation and possible abscess formation can result in tongue displacement leading to asphyxiation.
  - Alcoholics are more commonly affected.
  - Diabetics are more commonly affected.
- CT imaging is standard here.
  - If the clinical team suspects LA, your job is to search for clinically relevant pearls, no pun intended.
    - Mandibular odontogenic etiology, which is the most common cause, to include periapical root dehiscence.
    - Submandibular sialolithiasis.
    - Abscess formation.
    - Gas formation.
    - Degree of airway effacement.
  - If the clinical team is scanning for the sake of a Hail Mary, don't forget to assess these tissues. Look for edema. Look for fatty reticulation. Look for abscess. And report on the airway (Fig. 3.15).

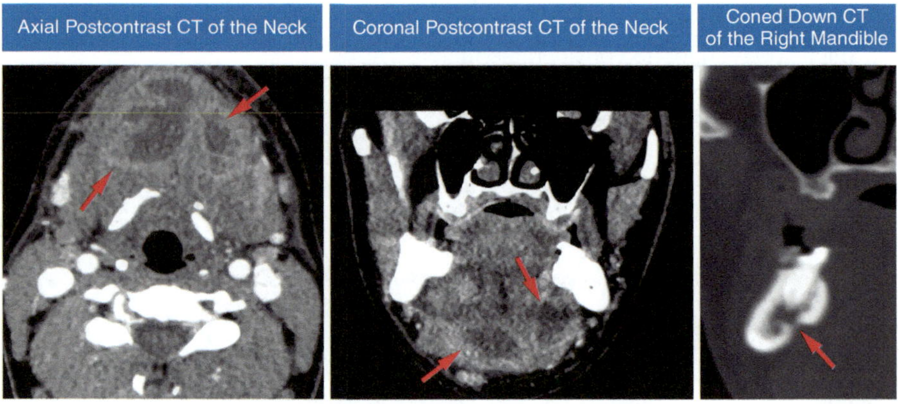

Fig. 3.15 The left and center annotated CT images reveal a multiloculated abscess within the submandibular and submental spaces. Recall that these spaces are in free communication with each other and are bounded superiorly by the mylohyoid muscle. The annotated CT image on your right shows the cause: a mandibular molar periapical abscess that eroded through the lingual cortex of the mandible. This erosion allowed direct purulent periapical communication with the submandibular space

3.4 Fungal/Atypical Infection

There are certain organisms that a neuroradiologist must know in order to be safe and helpful. Some of these organisms also overlap with popular test questions or *boogeymen* that frightened clinicians will query about every once in a while. For example, Naegleria *fowleri* is extraordinarily rare, but it's a media darling because it makes big headlines when a poor victim succumbs to it. *Brain-eating amoeba.* Has a macabre ring to it. On the other hand, rhinocerebral mucormycosis, an equally BAMF (as Samuel L. Jackson may say), is much more common and has a similar effect on its unfortunate diabetic victims. Anyway, on with the show.

3.4.1 Bottom Line Up Front

Some of the following entities may be considered common throughout the world but rare in the United States and vice versa. Some have specific imaging features, while most do not. A significant part of your radiologic acumen is possessing this knowledge. The ability to posit unusual diagnoses *in relevant and appropriate clinical scenarios* can mean the difference between delayed diagnosis, life, and death.

Five Diagnostic Pearls for the Resident/Fellow on Call:

- When in doubt, find the T-cell count.
- Know your locale and your "audience." A spelunker in Ohio will have different exposures than a missionary in Papua New Guinea.

- Query if the patient is on an immunosuppressive biologic medication, steroids, or a transplant recipient before you formulate a differential diagnosis.
- Prion illnesses are transmissible, and occasionally, you may come across an MRI that suggests a possible diagnosis of such. It is good etiquette to phone the patient's care team and tell them this. Extra precautions are needed.
- The differential diagnosis is your friend.

1. Tuberculous Meningoencephalitis (TBM)

   Tuberculosis has been a thorn in humanity's ass for millennia, with evidence of mycobacteria infections found in the ancient remains of Egyptian mummies [62]. No doubt massive tomes could (and have) been written about this disease, as it terrorizes humanity writ large in a most diverse manner. It is able to infect essentially every organ system we have, lie quiescent for decades, and evade/exploit our otherwise-awesome immunological defenses. Therefore, it is no surprise that it can cause trouble in the CNS.

   - TBM is a disease of young children and the immunocompromised, with HIV patients being at exceptional risk [63].
   - It is uniformly fatal if untreated.
   - Infection of the subarachnoid space by the organism can lead to ischemia and hydrocephalus.
   - MRI findings can be helpful as it has some predilections [64, 65]:
     - MRI findings will be present in a significant majority of patients with TBM.
     - Contrast enhancement in the basal meninges and Sylvian fissures is the most common finding.
     - Tuberculomas can be seen scattered throughout the brain parenchyma.
     - Ischemic infarcts may be seen in about half. Vessel wall imaging can show smooth enhancement, suggesting tubercular vasculitis.
     - Hydrocephalus is common (Figs. 3.16 and 3.17).
2. Cryptococcal Meningoencephalitis (CM)

   A disease of both the immunocompetent and immunocompromised, CM is a common and deadly manifestation of this fungus. About half of patients with CM will have HIV, while the remainder will be immunocompetent, and around 30% will die regardless of therapy [66]. Interestingly, pigeons are known to vector the disease, and one study showed geographical correlation between CM case density and pigeons [67].

   - *Cryptococcus neoformans* and *gattii* are the two main causative species.
   - CT findings are often absent while MRI findings may vary depending on whether the patient is HIV positive or not [68, 69].
     - Meningeal enhancement is more often present in patients who are immunocompetent.
     - Ependymal and choroid plexus enhancement is more common in immunocompetent patients.

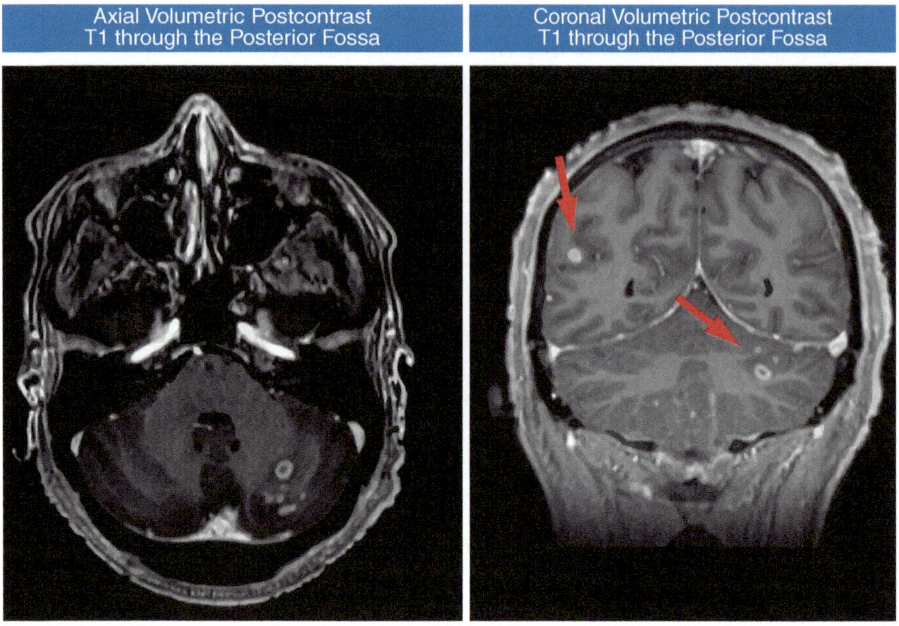

Fig. 3.16 *Mycobacterium tuberculosis* is a wily organism. With respect to the CNS, it can have multiple manifestations. Here we see speckled and ring-enhancing lesions in the supratentorial and infratentorial compartments. These turned out to be tuberculomas. The "textbook" basilar meningeal enhancement seen in other cases was not present here

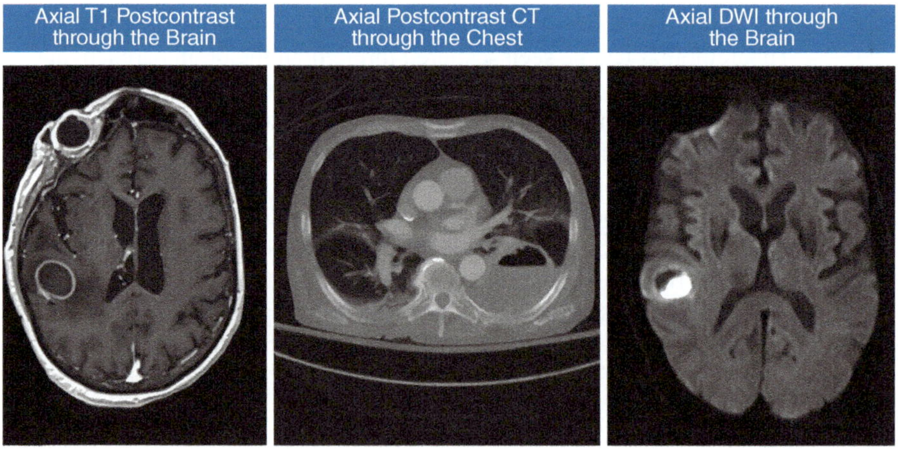

Fig. 3.17 In this series of images, we see a ring-enhancing mass in the right temporal lobe. This lesion partially restricts diffusion. The middle CT image shows a cavitary mass in the left lung. Another case of tuberculosis? It should be on your differential. However, this was a case of streptococcus bacteremia with lung and cerebral abscesses

  - Dilated perivascular spaces occur frequently in the basal ganglia in CM as they fill with gelatinous pseudocysts.
  - About half of CTs will be normal.
  - Infarctions in the bilateral basal ganglia are common.

3. Toxoplasmosis

   Caused by toxoplasma gondii, this protozoan parasite is number one of the top three opportunistic CNS infections in HIV positive patients who are not on prophylaxis. Though it infects about 1/3 of the world's population, it remains indolent in the immunocompetent. Whereas *Cryptococcus* is an infection associated with birds, toxoplasmosis is one associated with cats. And eating undercooked meat. Like a cat. Meow. The magic CD4+ T-cell count number at which this pathogen may reactivate is 100 cells/microliter or less.

   - This is often a subacute infection with nonspecific symptoms at onset.
   - The diagnostic accuracy of serum and CSF titers as well as PCR is low making this disease challenging to diagnose [70].
   - MRI and CT findings are nonspecific and overlap with other nasty pathologic processes. Other imaging modalities may be helpful [71].
     - CT scans may demonstrate low attenuating lesions within the basal ganglia, thalami, or at the cortical gray-white junctions.
     - MRI findings of toxoplasmosis overlap significantly with CNS lymphoma. Differentiating the two is very important for treatment.
       - Both can show ring enhancement.
       - Both can restrict diffusion.
       - CNS lymphoma demonstrates higher rCBV values on MR perfusion imaging (Fig. 3.18).
     - Thallium-201 nuclear medicine imaging has very high sensitivity and specificity for the purposes of differentiating these lesions. **Thallium-201 will concentrate in lymphoma.**

4. Rhinocerebral Mucormycosis

   Here we have a disease with a predilection for diabetics suffering from ketoacidosis but others with immunosuppression are vulnerable too. Interestingly, patients with elevated serum iron levels are at particular risk [72, 73].

   The infectious agents belong to the order Mucorales, which are commonly found in soil because *of course they are*. Infection begins in the paranasal sinuses and may be difficult to distinguish from regular acute sinusitis. However, this baddie wants more than just the paranasal sinuses. It rapidly invades the palate, mouth, orbit, and deeper structures. A black necrotic eschar is a telltale sign of mucor infection but only presents in 50% of cases [74]. Tissue necrosis is rapid. By the time it reaches the cavernous sinus (an advanced disease feature), mortality approaches 75% [75, 76].

   - There are clues on both CT and MRI that may suggest this diagnosis [77].
     - The patient's consolidated sinuses may demonstrate CT hyperdensity indicating fungal metabolism.

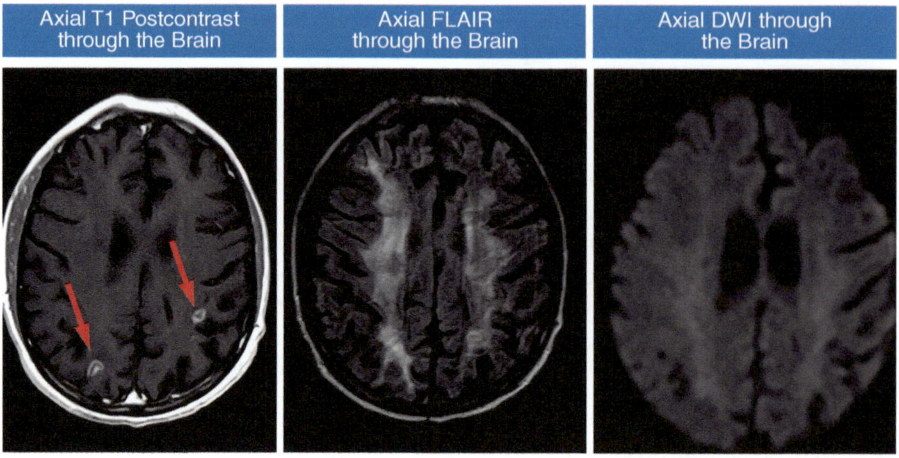

Fig. 3.18 The annotated axial MR image on your left shows some small irregular ring-enhancing masses. If I can challenge your credulity for a moment, does not the lesion in the patient's right brain parenchyma appear to have a more internal nodular component? A FLAIR image in the same patient shows cerebral atrophy and confluent FLAIR signal hyperintensities consistent with known HIV encephalopathy. DWI sequencing reveals no restricted diffusion of either of the ring-enhancing lesions. This turned out to be toxoplasmosis, but CNS lymphoma should be in your differential

   - Look for lytic and bone-permeative appearance of the sinus walls. This is suggestive of an aggressive process.
   - On MRI, T2 hypointensity in a paranasal sinus doesn't just mean air. It can also mean fungal colonization/consolidation.
   - On postcontrast MR imaging, look for the "black turbinate" sign in which lamina papyracea and turbinates are non-enhancing while the surrounding normal mucosa enhances. This indicates necrotic tissue.
   - Pre- and postseptal cellulitis are possible. Postseptal cellulitis can cause proptosis and strain the optic nerve.
   - Look for periantral (retromaxillary) fat effacement on precontrast T1 imaging.
   - Perineural invasion along the trigeminal nerve and through the cavernous sinus will manifest as avid enhancement (Figs. 3.19 and 3.20).

5. Prion Diseases

   Proteinaceous infectious particle (PREE-on) is the term for a very weird and frightening entity. Years ago, the diseases they caused were thought to be secondary to a "slow virus." In the interval since, scientists have discovered much about them. First, they are not viruses. They are rogue post-translational proteins that have been transformed from alpha configurations to beta-pleated sheets, which is the sole reason for their lethality [78]. Second, once transformed, they engender similar changes in wild-type proteins around them, which becomes a death (literally) spiral. Finally, not much is known about the wild-type function of the proteins, but they are ubiquitous in the human CNS.

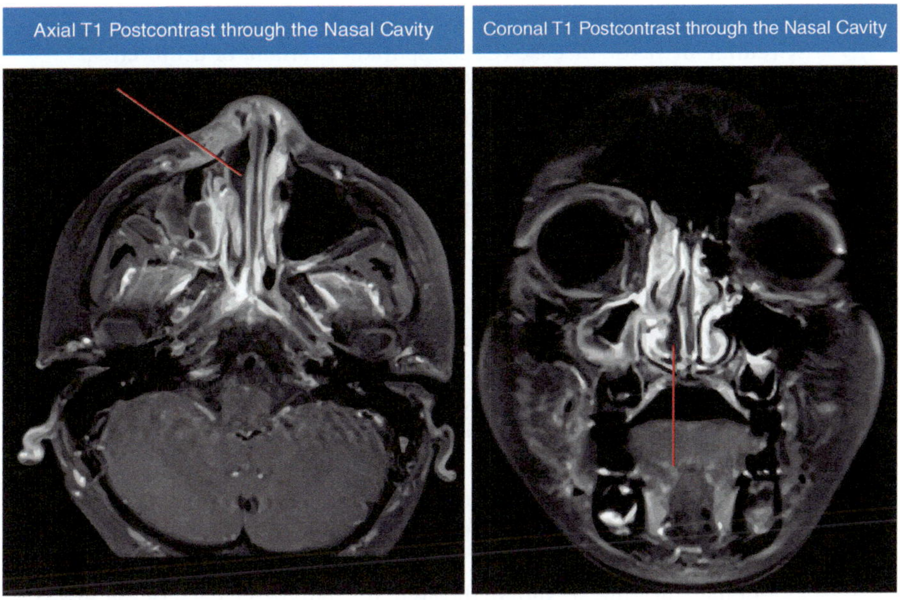

Fig. 3.19 A young immunosuppressed patient presented with copious nasal discharge and headache. Annotated T1 postcontrast multiplanar MR imaging is presented for your review. Note the mucosal thickening in the right maxillary sinus. However, this is not ordinary sinusitis. Pay attention to the inferior right concha and turbinate (red line). Part of it is **not** enhancing. It's devitalized and necrotic and thus non-enhancing due to invasive mucormycosis

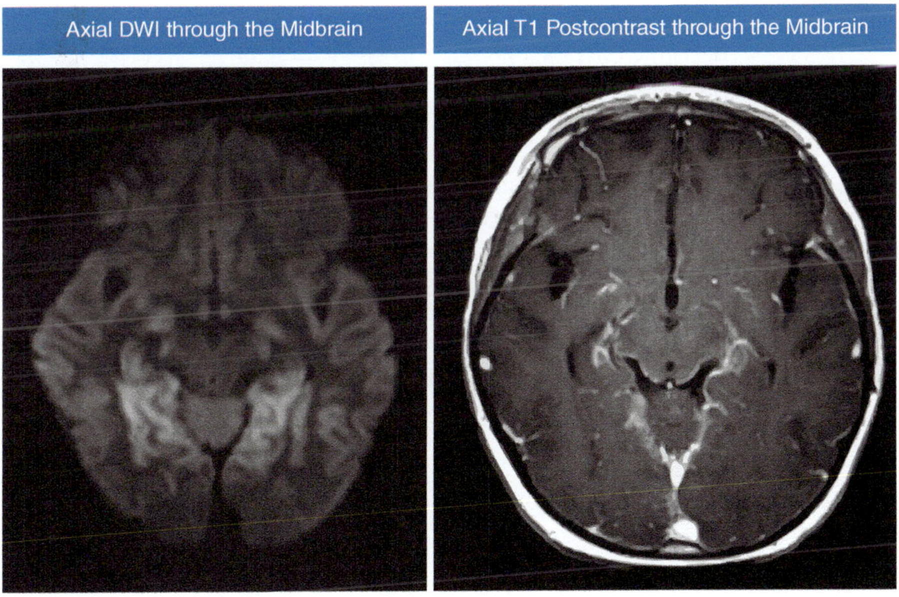

Fig. 3.20 This was a particularly horrifying case of mucor in a pediatric patient undergoing chemotherapy. It began as an aggressive sinus infection and ended with CNS dissemination and death. The image on your left shows gyriform restricted diffusion with a posterior preponderance. Axial T1 postcontrast imaging on your right reveals posterior basilar leptomeningeal enhancement

- The misfolded proteins cause specific diseases.
  - Creutzfeldt-Jakob disease (CJD): sporadic, variant, or acquired.
  - Gerstmann-Sträussler-Scheinker syndrome: familial.
  - Fatal familial insomnia syndrome: familial, autosomal dominant.
  - **All prion diseases are universally fatal.**
- Kuru is another acquired prion disease, which was discovered in Papua New Guinea and found to be the result of ritualistic cannibalism among the Fore tribe [79]. Kuru means "trembling" and is derived from the Fore language.
- Prion diseases are also known as transmissible spongiform encephalopathies (TSE) because they literally make the brain look like a sponge on autopsy.
- Prion diseases can be sporadic, iatrogenic, or acquired secondary to consuming infected material.
- MR imaging can help clinch the diagnosis of CJD with very high sensitivity, regardless of the etiology, but definitive diagnosis is postmortem [80, 81].
  - DWI signal hyperintensity is the most valuable indicator of CJD, and imaging findings must be taken in clinical context.
    - Single parietal cortex DWI signal hyperintensity is suspicious.
    - Single caudate DWI signal hyperintensity is suspicious. Look for additional curvilinear signal through the pulvinar. This is the so-called "hockey stick" sign.
    - Two DWI abnormalities in the cortical ribbons of the parietal and frontal lobes. Suspicious.
- Neuroimaging of fatal familial insomnia is very difficult because brain lesions are confined to the thalamus [82] (Fig. 3.21).

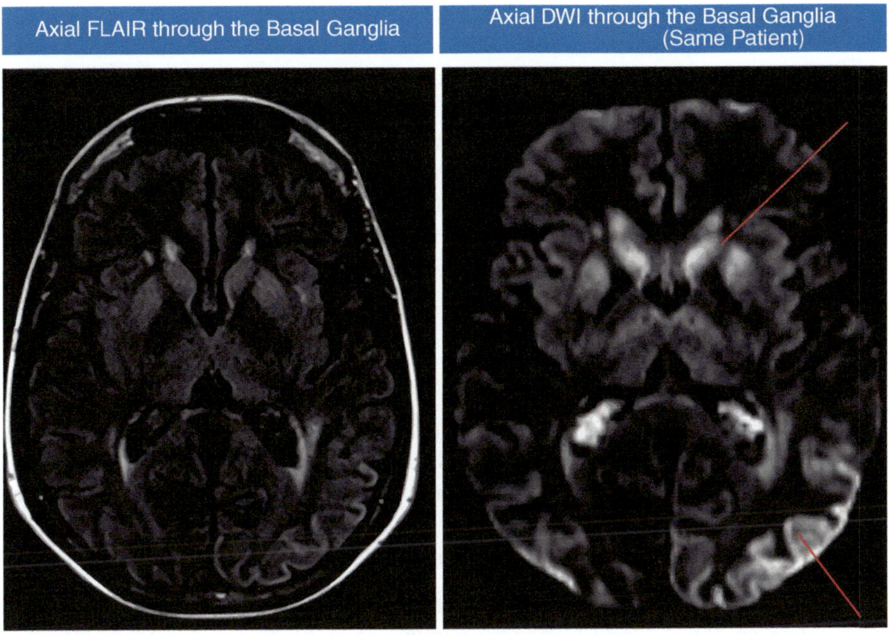

Fig. 3.21 Axial FLAIR imaging on your left reveals some FLAIR signal hyperintensity in the lentiform nuclei and caudate heads as well as the left occipital gyri. These findings are much more striking on axial DWI (right). In the setting of rapidly progressing dementia, these imaging findings are sensitive and specific for CJD

3.5 Demyelinating/Inflammatory/Autoimmune

When the immune system turns against a person, chaos and sorrow ensue. The manifestations of autoimmune disease are protean depending upon what the host's immune system decides to attack. Further, when modern medicine attempts to mitigate the effects of autoimmune disease, chaos and sorrow may *still* ensue. Sometimes. Not to denigrate the amazing progress science and medicine have made, but attempting to manage autoimmune disease is like trying to play a game of Jenga. From two miles away. Using a telescope and a rifle.

3.5.1 Bottom Line Up Front

You will frequently see CNS manifestations of autoimmune disease as well as primary demyelinating diseases. Many of these diseases are chronic but can have acute and devastating manifestations, which is why I include them here. I also realize this topic is expansive. In the interests of brevity, and as much as it pains me, only the most acutely relevant and diagnostically salient facts will be included.

Five Diagnostic Pearls for the Resident/Fellow on Call:

- Demyelinating diseases can masquerade as strokes clinically and the plaques can do the same on CT.
- When assessing the orbits on MRI, make sure you carefully examine the optic nerves retrograde through the optic chiasm and visualized optic tracts. The differential diagnosis will change depending on what is affected.
- Pre- and postcontrast imaging is exceptionally important in this setting.
- **SWI sequences are superior when evaluating for MS, which is reflected in the new (2024) McDonald Criteria.**
- Don't forget the spine.

1. Demyelinating Diseases

   Myelin is essential for normal propagation of neuronal signals along axons. It puts the white in white matter and has a complex composition. Unfortunately, it is especially vulnerable to both autoimmune attack and exogenous toxic effects (discussed later). Given that myelin is a vulnerable and complex material, it presents multiple targets for the immune system to attack.

   - **Multiple Sclerosis**: This is the best known and most common demyelinating disease. It has a predilection for patients aged 20–40 years and tends to affect women 2–4x more than men [83]. Moreover, the disease has latitudinal gradient, with a higher prevalence the farther one goes from the equator [84].
     - Diagnosis of this disease requires **dissemination in time and space**. These are important buzzwords. Remember them.
     - A combination of clinical and MRI findings is essential in the diagnosis. The 2024 McDonald Criteria describe this (Table 3.1).
     - MRI hallmarks of the disease are FLAIR signal hyperintensities in specific distributions. These distributions define dissemination in space. Dissemination in time is a bit more complex.
     - Look for **Dawson's fingers**. These are white matter FLAIR signal hyperintensities oriented perpendicularly along the corpus callosum.
     - **Optic neuritis is not specific for MS. It is more specific if it is short-segment and excludes the chiasm**.
     - Note that diagnosing MS on call is not one of your priorities. The table above is mostly for the purposes of quick reference and routine knowledge. However, providing good imaging descriptions of what you see greatly assists the clinicians arrive at a diagnosis (Figs. 3.22, 3.23, 3.24, 3.25, and 3.26).
   - **Acute Disseminated Encephalomyelitis (ADEM)**

     One can think of ADEM as MS's frenetic and impatient cousin. It is most often a monophasic disease that "visits" after a viral illness or rarely vaccination, though these points are still contentious as the etiology remains poorly understood [85]. Unlike MS, ADEM strikes younger patients and tends to do so during the winter and spring. It also has no gender predilection. The

Table 3.1 The 2024 McDonald Criteria for the diagnosis of MS distilled into a table. Dissemination in space (DiS). Dissemination in time (DiT)

2024 McDonald Criteria for the diagnosis of multiple sclerosis	
Dissemination in space	Dissemination in time
Periventricular, juxtacortical, Infratentorial, spinal cord, optic nerve (must have >1 lesion in at least 2 of these territories)	Enhancing and non-enhancing white matter lesions fulfills this criterion. If all are non-enhancing, a follow-up exam is necessary. A new lesion must appear to fulfill DiT
Lesions in 4 of the 5 territories does not require DiT or CSF analysis. MRI is diagnostic in this setting	Oligoclonal bands or kappa-free light chains in CSF analysis satisfy the DiT criterion
Additional diagnostic imaging findings (new from 2017)	
Central vein sign: SWI signal hyperintensity surrounding a thin hypointense vein less than 2 mm in thickness Need at least 6 lesions in 1 typical location + DiT	Paramagnetic rim sign: Thin black SWI line surrounding a lesion. Need at least 1 lesion in 1 typical location + DiT
Special situations	
Primary progressive MS > 12 months ➔	2 spinal cord lesion = DiS
No clinical signs/symptoms of MS Radiologically isolated syndrome (RIS) ➔	DiS + DiT or DiS + oligoclonal bands or DiS + 6 central vein sign lesions

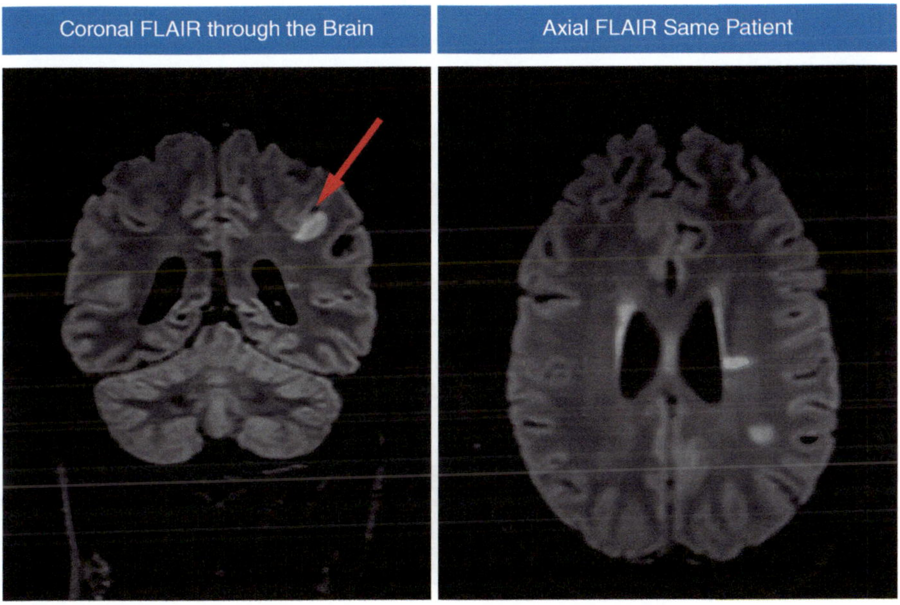

Fig. 3.22 Coronal and axial FLAIR imaging through the brain reveals FLAIR signal hyperintensities that are suspicious for demyelinating disease. The annotated coronal image on your left shows a juxtacortical lesion. Juxtacortical means it touches the gray matter with no intervening white matter. The axial image on your right reveals periventricular FLAIR signal hyperintensity that is perpendicular to the ventricle. Similarly, periventricular means it touches the ependymal surface with no intervening white matter. These findings are common in MS

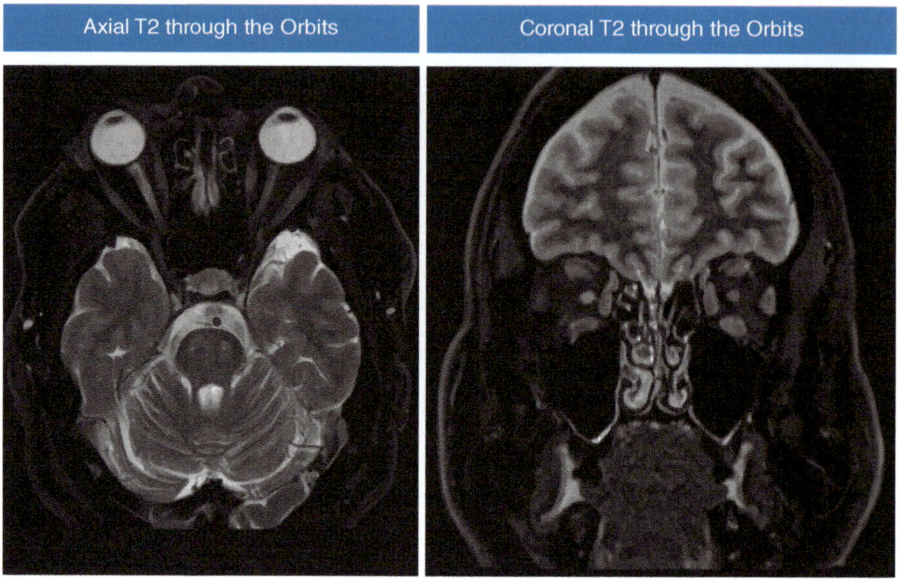

Fig. 3.23 Axial T2 (your left) and coronal T2 (your right) images show asymmetrically increased signal through the entire length of the patient's right optic nerve. This finding can be subtle in some patients, so window/level the scan carefully and scrutinize for symmetry. It is not subtle in this case. Plot twist: this case was actually *not* one of MS. Read on

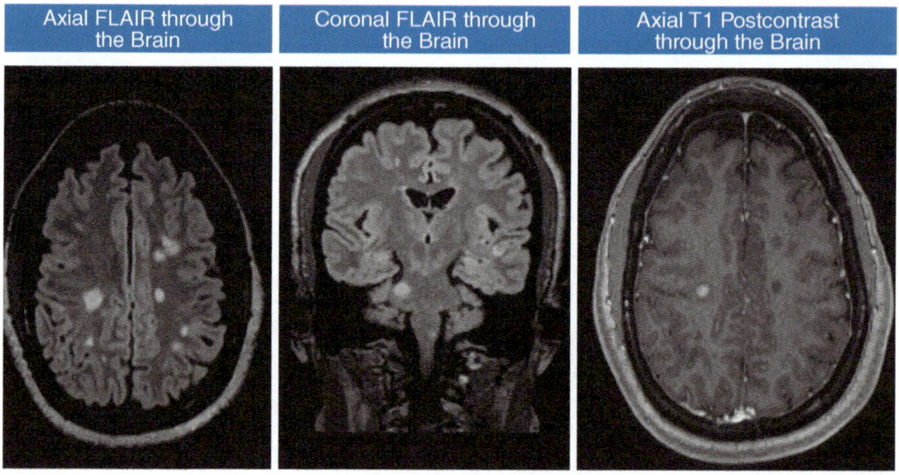

Fig. 3.24 This triad of MR images shows a "packaged" diagnosis of MS. There are periventricular, juxtacortical, and infratentorial white matter FLAIR signal hyperintensities. One of the visualized lesions in the patient's right corona radiata enhances, whereas the others do not. The requirement of dissemination in time and space is fulfilled

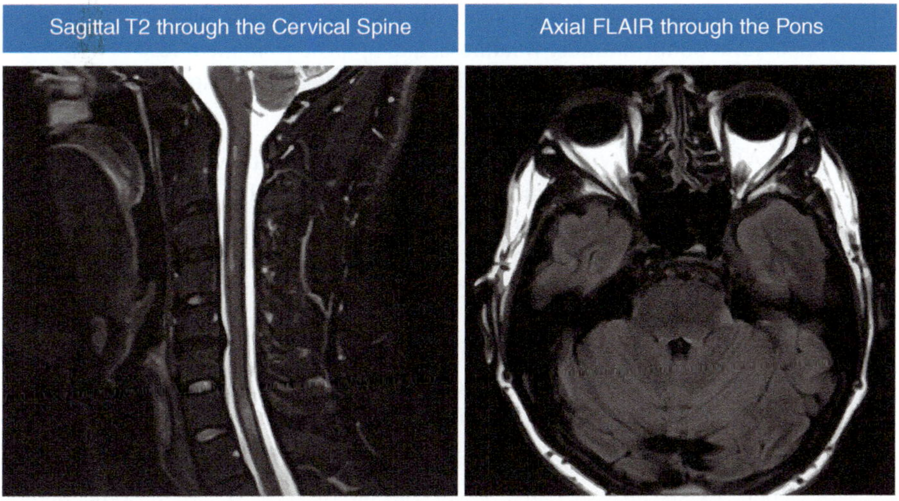

Fig. 3.25 The image on your left reveals short segment T2 signal hyperintensities at multiple levels in the cervical spinal cord. The right image shows FLAIR signal hyperintensity in the patient's left trigeminal root entry zone, which is frequently seen in MS. This can be a cause of trigeminal neuralgia

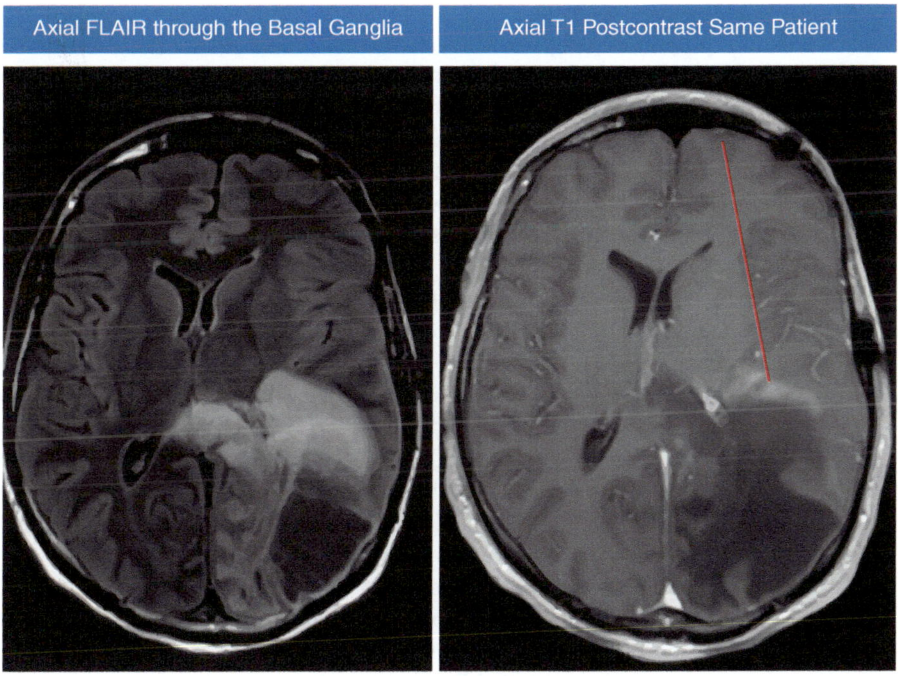

Fig. 3.26 Tumefactive MS—a challenging and perplexing entity. Notice the mass effect and FLAIR signal hyperintensity involving the splenium of the corpus callosum seen on the axial FLAIR image on your left. This entity can mimic glioblastoma. However, the annotated image on your right (red line) shows partial peripheral enhancement, a clue that it's actually a demyelinating process. "Partial rings demyelinate things"

clinical presentation is also more diverse with fever, visual field defects, behavioral abnormalities, weakness, and aphasia to name a few.

- MRI findings can be quite similar to MS, but there are some discriminating clues [85–87]:
  - ADEM will typically spare the periventricular white matter and one will *not* see perpendicular white matter FLAIR signal lesions to the ventricles/corpus callosum (Dawson's fingers).
  - ADEM can involve the gray matter, especially the deep gray nuclei.
  - T1 hypointensities ("black holes") are less common in ADEM.
  - Spinal cord lesions can be short or longitudinally extensive (greater than three vertebral segments in length).
  - Brainstem involvement is possible with one study suggesting midbrain involvement is more common in adults and pons involvement more common in children.
- There exists an especially nasty variant of ADEM called acute hemorrhagic leukoencephalitis (AHLE) also known as Weston-Hurst disease. The clinical presentation is hyperacute and much worse than ADEM. The prognosis is most often fatal as well.
  - MRI may show a greater burden of brain lesions, including tumefactive ones. Gradient and SWI can show lesional hemorrhage [88] (Fig. 3.27).

- **Guillain-Barré Syndrome**
- GBS is a polyneuropathy with an increasing global prevalence. Similar to ADEM, GBS occurs after a viral or bacterial infection, eliciting an unfortunate episode of molecular mimicry wherein the immune system bashes peripheral nerves. Interestingly, this disease is most strongly associated with prior *Campylobacter jejuni* infection [89]. It is an ascending paralytic monophasic process that begins in the lower extremities and climbs, also affecting the respiratory muscles and bulbar system. Autonomic dysfunction is possible as well, which can be fatal [90].
  - MRI of GBS can show avid enhancement of the conus medullaris and cauda equina with involvement of the ventral nerve roots more specific to the disease [91].
  - Cranial nerve enhancement is possible too with the facial nerve being most commonly affected.
  - The MR imaging and clinical findings can overlap with chronic inflammatory demyelinating polyneuropathy (CIDP) during an acute episode of this disease. Other differential considerations include leptomeningeal carcinomatosis, neurosarcoidosis, and arachnoiditis [91, 92] (Fig. 3.28).
- **Myelin Oligodendrocyte Glycoprotein-Associated Disease (MOGAD)**

  I mentioned that myelin is a complex and vulnerable protein. Part of that reasoning is because it has a lot of constituents that are vulnerable to the immune system. Indeed, MOGAD is a monophasic *or* relapsing process, and the result of anti-MOG antibodies that damage myelin in predictable distributions. Being a relatively new player on the stage, this disease only recently

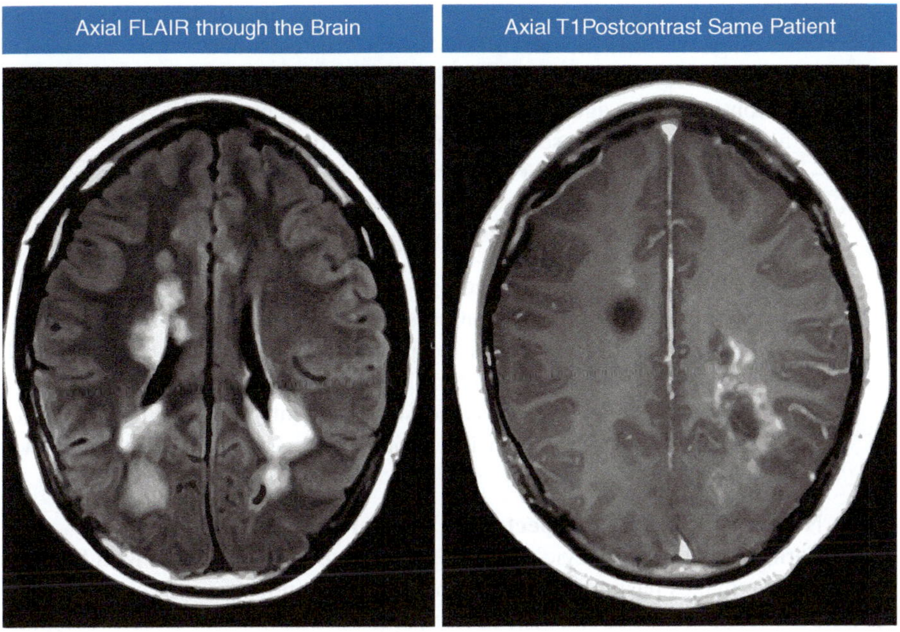

Fig. 3.27 Here is a case of ADEM in a young patient who was recovering from an unremarkable upper respiratory infection. The FLAIR signal hyperintensities are globular and periventricular. Some of them demonstrate patchy peripheral contrast enhancement. The T1 hypointensity in the right corona radiata is indeed a "black hole." This case breaks the rules described above. That's radiology

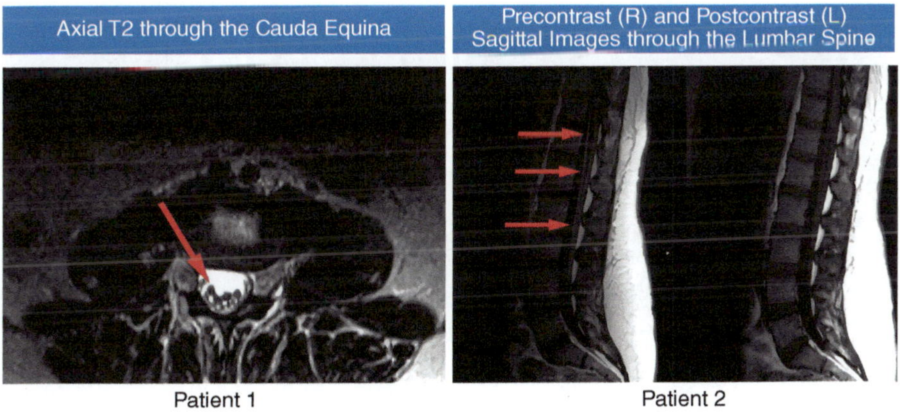

Fig. 3.28 The annotated axial T2 image through the cauda equina reveals ventral nerve root thickening in Patient 1. Patient 2 shows avid enhancement of the cauda equina as annotated by red arrows. Both patients suffered from ascending paralysis consistent with GBS

(2023) received a proposed consensus diagnostic mechanism [93]. In short, other diagnoses like MS and NMO-SD must be excluded.

- Unlike MS, MOGAD has no age or sex predilection. However, its presentation varies by age with an ADEM-like presentation in young children and optico-spinal presentation in older children and adults [94].
- Optic neuritis is the most common onset phenotype in adults [95].
- MRI findings in MOGAD overlap with other demyelinating disease processes and also depend heavily on age [96, 97].
  - Pediatric patients show findings similar to ADEM.
  - About one third of adult patients will have brainstem involvement.
  - MOGAD optic neuritis is longitudinally extensive but preferentially affects the anterior segment of the optic nerve.
  - Transverse myelitis in this disease is frequently H-shaped (gray matter) and longitudinally extensive, though it can be short segment as well (Figs. 3.29 and 3.30).

- **Neuromyelitis Optica Spectrum Disorder (NMO-SD)**
  Formerly known as Devic's disease, NMO was later expanded with SD. This disease makes hashtags out of astrocytes when antibodies target the aquaporin 4 (AQP-4) water channel. Indeed, NMO-SD is more of an astrocytopathy than a demyelinating disease [98]. However, I find it helpful to group it with better known demyelinating disease processes in order to recall the MR imaging associations. As an aside, in 2005 I had the distinct pleasure of meeting Dr. Peter Agre, one of the discoverers of aquaporin water channels. Definitely

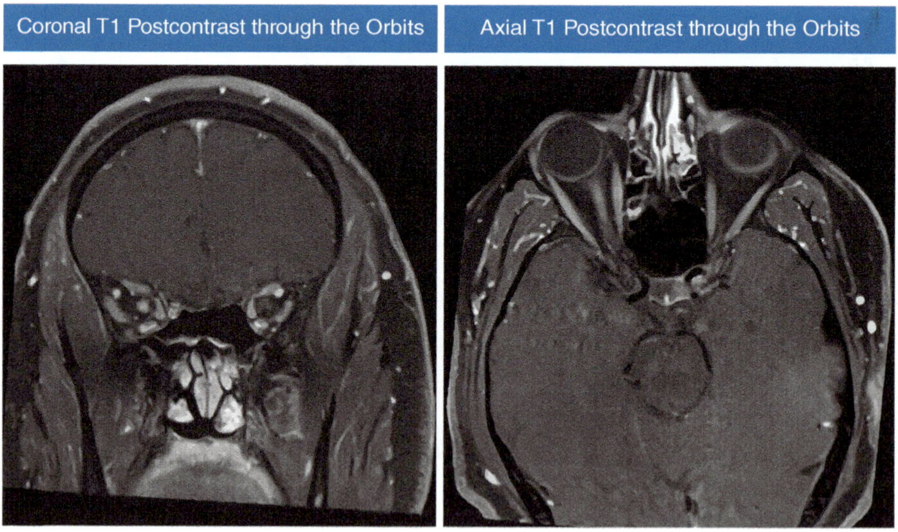

Fig. 3.29 Multiplanar postcontrast fat-saturated T1 imaging through the orbits shows extensive enhancement of the patient's intraconal right optic nerve. There was sparing of the optic chiasm. This finding is more common in MOGAD than in MS

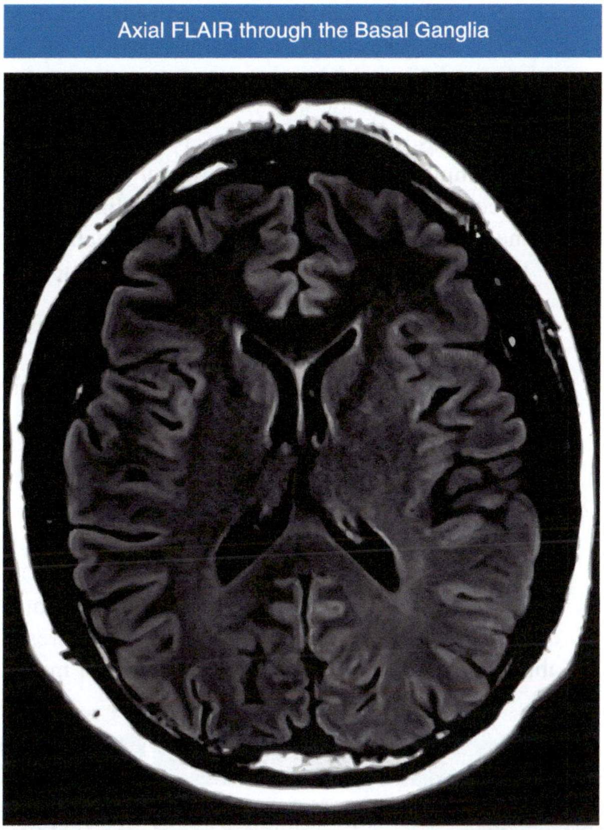

Fig. 3.30 This patient's brain appeared normal on MR imaging. Long segment optic nerve enhancement was the only presenting radiologic finding of MOGAD

a humble and gracious man. I wonder if, at that time, he knew of its implications in clinical medicine.

- NMO-SD has a higher prevalence among Blacks and East Asians with an extraordinary female preponderance at a ratio of 9:1 [99].
- A long time ago, in a chapter far, far away, I mentioned intractable hiccups as a symptom of Wallenberg Syndrome. Also recall that it wasn't pathognomonic for lateral medullary stroke, and that intractable hiccups may also be seen in NMO-SD.
- MR imaging findings in NMO-SD are variable (aren't they all?), but there are some useful tidbits to know, which also help explain the hiccups [100, 101]!
    - Involvement of the optic nerves is common but NMO-SD more often involves the posterior segments, including the chiasm, and it is more often bilateral.
    - Commensurate with its watery aquaporin associations, NMO-SD can attack the dorsal brainstem, irritating the area postrema at the floor of the fourth ventricle, and the tractus solitarius. This is a more specific finding and associated with nausea, vomiting, and hiccups.

   - NMO lesions also tend to follow the ependymal lining in the lateral ventricles (Figs. 3.31 and 3.32).

- **TNF-Alpha Blocker-Associated Demyelinating Disorder**

  This class of drugs has been a blessing for many people who suffer from autoimmune disease. As with many drugs, there exists a heterogeneous side effect profile, including CNS demyelination, though this is quite rare. The mechanism of action for CNS demyelination is not known but withdrawal of the offending drug usually leads to resolution [102–104].

  – It can present with various neurologic symptoms that are nonspecific, including paresthesia and ataxia.
  – In many patients, the demyelination is asymptomatic and incidentally discovered (Fig. 3.33).
  – The MRI findings can mimic NMO-SD.
    - Transverse myelitis has been reported.

2. Inflammatory and Autoimmune Disease

   I have selected a couple diseases to include in this section in the interests of brevity. Be aware that, generally speaking, any autoimmune disease can have CNS manifestations. Also recognize the fact that you *cannot* package up every disease process with a nice bow in your impression. Your job is to recognize abnormal and describe it thoroughly, but in many (if not most) instances, you

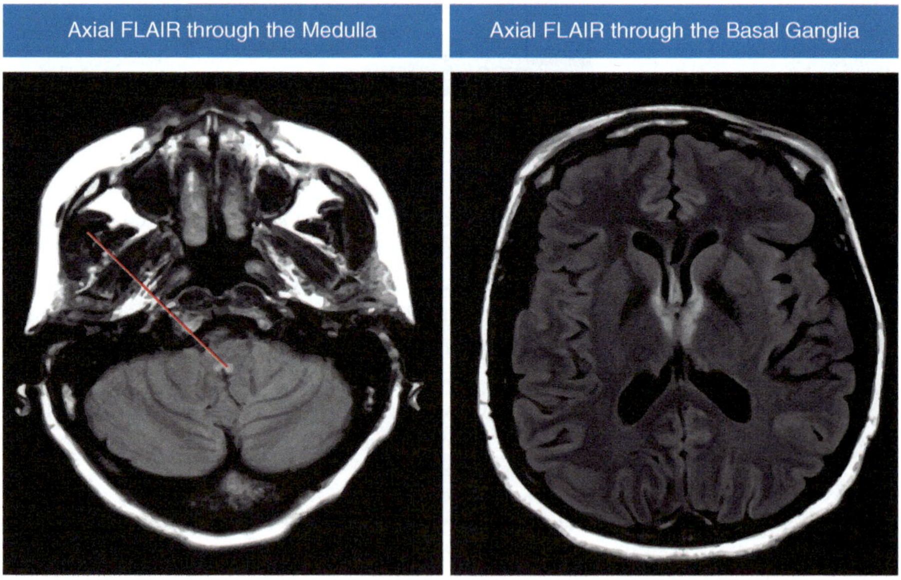

Fig. 3.31 The image on your left is annotated and points at FLAIR signal hyperintensity within the dorsal periventricular medulla in the vicinity of the area postrema. The axial FLAIR sequence on your right reveals abnormal FLAIR signal along the periventricular diencephalon. These are aquaporin rich areas

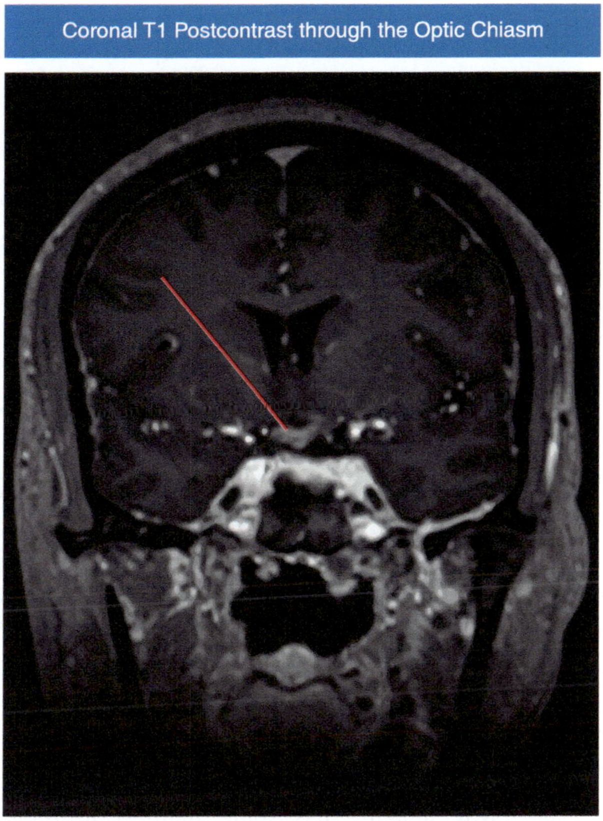

Fig. 3.32 Coronal T1 postcontrast fat saturated imaging with annotation shows strong enhancement of the optic chiasm. This is more common in NMOSD

cannot say for certain that it is *disease x*. Attempting to do so can have huge repercussions for you, the referring clinician, and even the patient if their insurance company gets ahold of the report. Use your words judiciously. Many of the following diseases will exhibit a perivascular enhancement pattern on MRI. I highly recommend reading the Babin et al. article in the Journal of Neuroradiology, which discusses in detail the findings and differential diagnoses [105].

- **Susac's Syndrome**

  A rare disease of the senses, Susac's syndrome is an immune-mediated microangiopathy that results in sensorineural hearing loss, retinal artery occlusion with vision loss, and encephalopathy [106]. It is predominantly (80%) a disease of young women [107]. Another eponymous disease with a similar clinical picture is Cogan syndrome though this disease does not demonstrate corpus callosum involvement.

  - The disease has a "traditional" triad that doesn't manifest all at once but does occur in most patients over the long term.
  - Behavioral changes and focal neurologic deficits occur. These include problems concentrating, psychosis, ataxia, and aphasia to name a few.

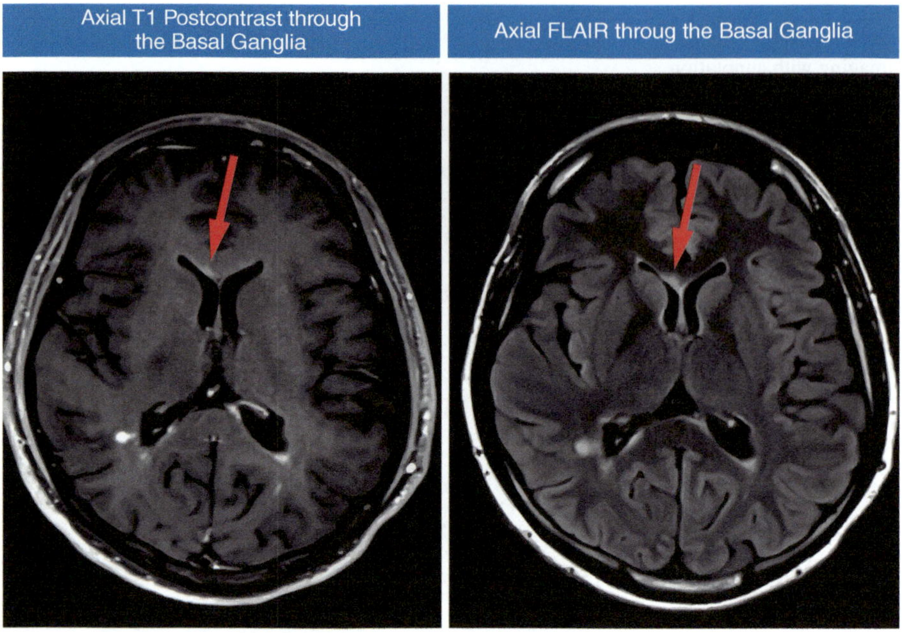

Fig. 3.33 Imaging from a patient with Crohn's disease who began a TNF-alpha blocking medication. The patient developed multiple motor and sensory symptoms, which precipitated this MRI of the brain. The annotated images show FLAIR signal hyperintensities that also enhance. After the medication was stopped, these lesions resolved over time

- MR imaging has some findings useful for diagnosing this disease [108, 109].
  - Original descriptions include a triad of signal abnormality in the corpus callosum, deep gray structures, and leptomeningeal enhancement.
  - The callosal lesions are rounded in shape, described as "snowball" and localized to the body.
  - Enhancement of the cochlea and semicircular canals is possible.
  - Black blood vessel wall imaging can also be helpful to demonstrate a small vessel vasculitis. Look for small vessel and perivascular enhancement.

- **Neurosarcoidosis**
  Non-caseating granulomas! Buzzwords you may recall from your medical student days. Sarcoidosis is another multisystem inflammatory disease that your chest/thoracic colleagues are most familiar with because the lungs are involved in 90% of cases. Neurologic involvement, in contrast, is between 5% and 10% [110]. Common presentations include meningitis, cranial neuropathy, myelitis, and pituitary disease resulting in neuroendocrine dysfunction.

  - Sarcoidosis is a disease of unknown etiology that can affect anybody but the highest prevalence is seen in Black Americans and people of North African ancestry [111].
  - MRI findings are protean and overlap with other disease processes [112–114].
    - Thick nodular meningeal enhancement is common on MRI and can mimic tuberculous meningoencephalitis.
    - Cranial nerve enhancement is common but does not correlate well with clinical symptoms. Cranial neuropathy is most common in nerves 2, 7, and 8.
    - Intraparenchymal mass-like enhancement is possible, which can mimic a primary CNS neoplasm.
    - There can be enhancement of the pituitary gland and infundibulum.
    - Communicating hydrocephalus with transependymal flow is possible (Fig. 3.34).

- **Neuro-Behçet's Disease (NBD)**
  A non-discerning vasculitis, Behçet's disease affects vessels of all sizes, including arteries and veins. The pathophysiology of this disease is not well understood, and anecdotally it is absolutely miserable. The manifestations are

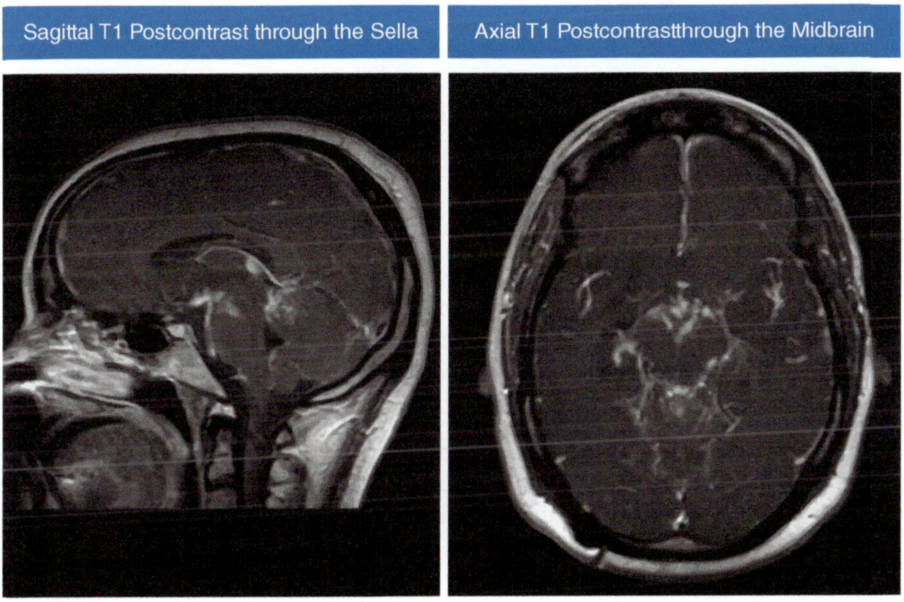

Fig. 3.34 The sagittal T1 postcontrast image on your left shows thick nodular enhancement along the surfaces of the ventral pons, interpeduncular cistern, mammillary bodies, tuber cinereum, and anterior superior cerebellum. Axial T1 postcontrast imaging of the same patient typifies basal cistern enhancement as well as meningeal enhancement along the interhemispheric fissure. This was a case of neurosarcoidosis, though tuberculous meningitis can look very similar

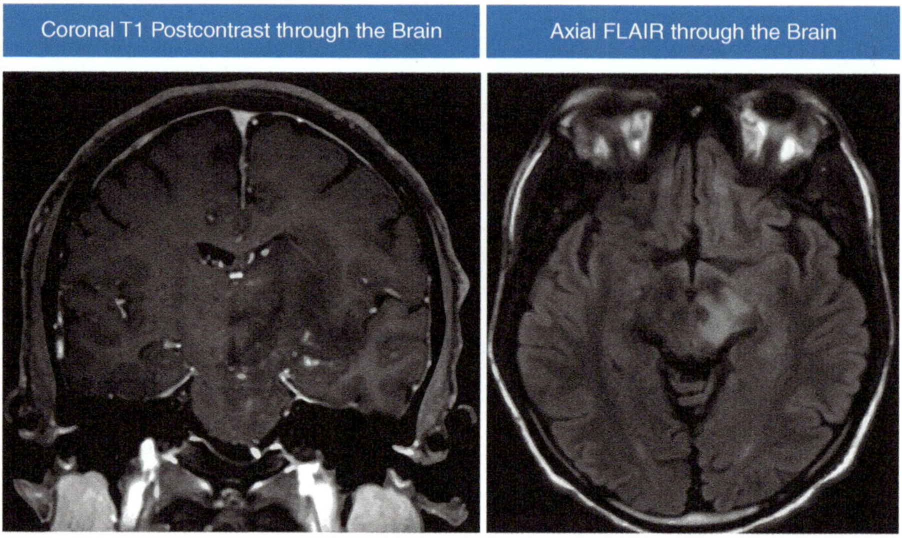

Fig. 3.35 Coronal T1 postcontrast fat saturated imaging on your left reveals edema, mass effect, and scant enhancement that appears to ascend from the patient's left brainstem up through the thalamus. The axial FLAIR image shows left midbrain FLAIR signal hyperintensity that involves the left cerebral peduncle but spares the red nucleus. The common descriptor here is "meso-diencephalic lesion." **Sparing of the red nucleus is an important finding as well**. The parenchymal distribution is striking in this disease. It's curvy like that little tail on the "ç" in the good Turkish doctor's name

mucocutaneous, including aphthous ulcers in the mouth, and genital lesions, though most bodily systems can be affected. **Between 10 and 30% of patients with Behçet's will suffer NBD with the most common symptom being migraine** [115]. Additional neurologic symptoms include pyramidal signs, dysarthria, and hemiparesis.

- It is also known as the Silk Road disease based on its demographics. Young Middle Eastern and Turkish men are most commonly affected.
- MRI is useful in both the parenchymal and extra-parenchymal manifestations of Behçet's, and there are certain patterns the radiologist should notice [116, 117] (Fig. 3.35).
    - Striking FLAIR/T2 signal abnormalities at the meso-diencephalic junction are hallmarks of this disease. Curvilinear signal "climbs" from the brainstem through the cerebral peduncle into the thalamus and internal capsule.
    - Brainstem lesions are usually ventral.
    - Extra-parenchymal involvement includes cranial nerve enhancement, intracranial hypertension, and dural venous sinus thrombosis.
    - Meningoencephalitis is possible with avid postcontrast enhancement of the meninges.

- **Primary Angiitis of the CNS (PACNS)**
  PACNS is a protean vasculitis and a difficult diagnosis to make. There are many vasculitides, each with its own vascular predilection. Medicine tends to categorize these diseases by the size of the vessels they affect. Some have imaging findings suggestive of their diagnosis while many do not. Indeed, the term vasculitis simply reflects inflammation of vessels. As we recall, inflammation can be immune mediated or infectious. For example, COVID-19 is an endotheliitis [118].
  - PACNS is considered a small vessel vasculitis and a disease that affects men twice as often as women. It has a mean age of onset of 50.
  - Clinical presentation is nonspecific and includes headache and encephalopathy and symptoms outside of the CNS are rare, which can confound the diagnosis.
  - Individuals with this disease are frequently and erroneously thought to have vasospastic disease such as RCVS [119]. Distinguishing between the two is essential due to treatment differences [120].
  - MRI abnormalities are present in 90% of patients with this disease but are nonspecific and digital subtraction angiography (DSA) appearance of PACNS is essentially the same as RCVS [121, 122] (Figs. 3.36, 3.37, and 3.38).

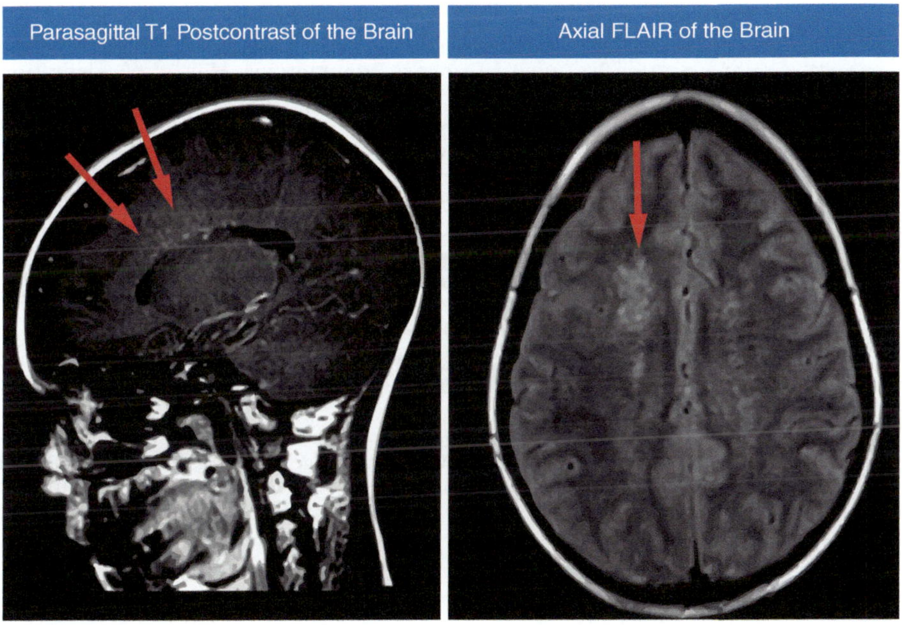

Fig. 3.36 Sagittal postcontrast T1 imaging on the left (arrows) reveals fine linear enhancement along the anterior periventricular right centrum semiovale. This finding is suggestive of a perivascular enhancement pattern. The annotated axial FLAIR on your right shows concomitant and overlapping FLAIR signal hyperintensity. The differential is broad

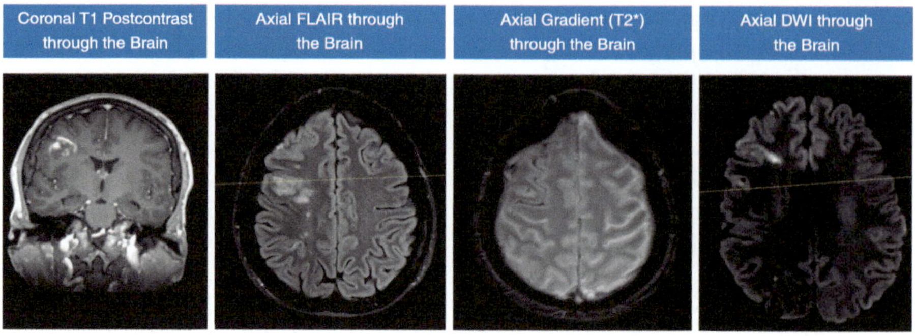

Fig. 3.37 Coronal T1 postcontrast image on your far left reveals nonspecific parenchymal enhancement. There is patchy FLAIR signal hyperintensity overlapping this enhancement. Gradient sequencing shows evidence of prior hemorrhage due to hemosiderin staining in the right superior frontal sulcus. DWI shows evidence of a superimposed acute infarct. This constellation of findings is very unusual in a 30-year-old patient

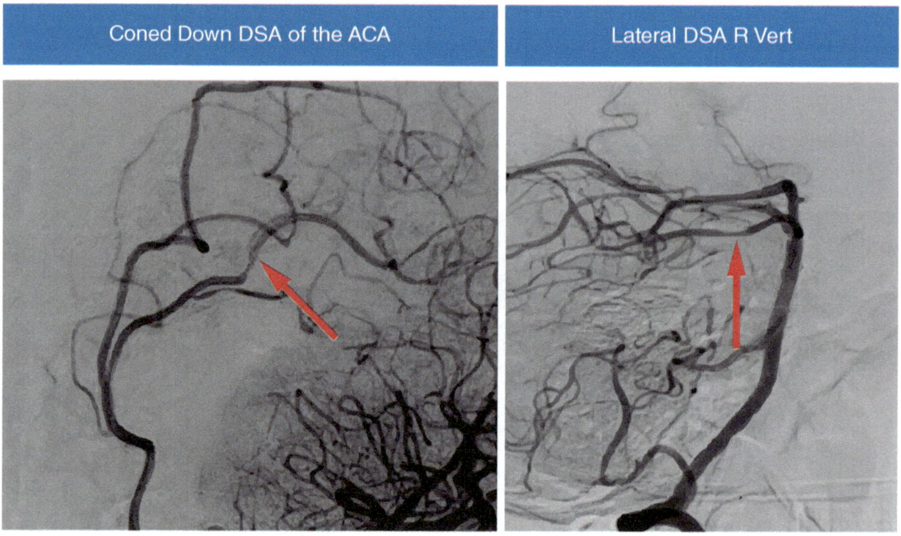

Fig. 3.38 DSA of the same patient depicted in 3.37 was obtained. The red arrows annotate stenoses within the distal ACA (left image) and the PCA (right image). The constellation of MRI and DSA findings is suggestive of PACNS

- Subcortical white matter FLAIR signal hyperintensities? Yep.
- Restricted diffusion in the setting of acute infarct? Yep.
- Microhemorrhages? Yep!
- Vessel wall imaging may be exceptionally helpful in this setting, demonstrating concentric vessel enhancement.

- **Immune Reconstitution Inflammatory Syndrome (IRIS)**
  A phenomenon that occurs after initiation of antiretroviral therapy, IRIS is the result of quick reconstitution of the immune system in patients with low CD4+ T-cell counts. It is more likely to occur in the setting of varying mycobacterial infections, herpes infection, Cryptococcal disease, PML, or CMV infection. The consequences can be devastating, especially in male patients. Additional risk factors for death include cryptococcal or TB meningitis, low BMI, and anemia [123, 124] (Fig. 3.39).

  – CNS IRIS is categorized with the infectious agent that is present at the time of immune reconstitution. For example:

    - PML-IRIS (JC virus)
    - CM-IRIS (*Cryptococcus*)

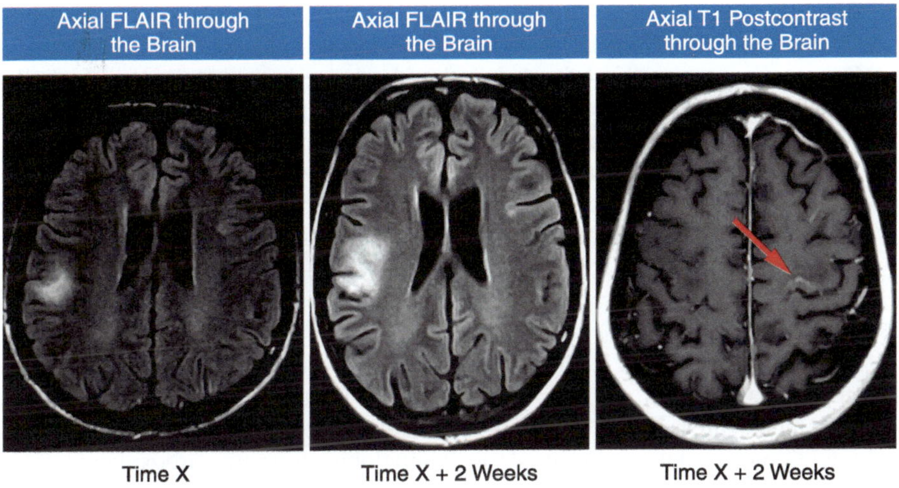

Fig. 3.39 The image on your left was obtained in a patient with an undetectable CD4+ count. The right subcortical FLAIR signal hyperintensity appears to involve the cortical U-fibers, suggesting PML. At that time the patient was started on antiretroviral therapy. Repeat imaging 2 weeks later reveals growth of the lesion and a contralateral new focus of parenchymal enhancement (arrow). These findings are consistent with PML-IRIS, an acute exacerbation of disease while on the road to recovery

3.6 Toxic Encephalopathy

The human brain is an organ of extraordinary complexity, mystique, and function. In other words, it is *very* high maintenance. As we've discussed, mere minutes of oxygen deprivation causes it to shut down, and damage sustained is often irreversible. But wait—there's more. The brain is also vulnerable to a huge number of exogenous chemicals, some of which human beings actively seek, while others are accidentally or surreptitiously consumed. We will discuss a few of these.

3.6.1 Bottom Line Up Front

You will see some these cases while on call because the patient is acutely ill. Your job is to recognize the pattern of illness and provide a differential. Some of these imaging patterns demonstrate significant overlap with not only other toxic encephalopathies but also metabolic ones. Other cases will present with an acute exacerbation of a chronic illness. Alcoholism immediately comes to mind.

Five Diagnostic Pearls for the Resident/Fellow on Call:

- As usual, the first examination you see will probably be a noncontrast head CT. Pay close attention to the basal ganglia and the internal capsules.
- Bilateral symmetric hypoattenuation of deep gray matter structures or white matter tracts should raise your suspicion for a toxic encephalopathy.
- Do not discount a toxic encephalopathy if you see brainstem and/or cerebellar involvement. Some of these processes will involve posterior fossa structures.
- All of these processes deserve a phone call to the referring clinician. Many can be treated or reversed with simple measures, and delayed action can be fatal.
- MRI diagnosis can be especially tricky in pediatric patients due to age-dependent myelination patterns that differ from a normal adult's MRI appearance.

1. Carbon Monoxide (CO)

   CO exposure is the leading cause of human poisoning worldwide. There are up to 50,000 cases of CO poisoning in the United States every year [125]. Common sources of CO exposure include improperly vented heaters, fires, and intentional inhalation. CO binds to hemoglobin, forming carboxyhemoglobin in vivo. Carboxyhemoglobin binds strongly to oxygen, preventing its release into peripheral tissues and causing tissue hypoxia. CO poisoning has variable presentation, dependent on the degree of CO exposure. Common symptoms of CO poisoning include headache, dizziness, fatigue, nausea/vomiting, altered mental status, chest pain, shortness of breath, and loss of consciousness. Severe acute CO poisoning manifests as cognitive dysfunction, which can quickly progress to progressive brain injury with edema. Neurological symptoms of CO poisoning may present immediately or as delayed sequelae occurring 2–6 weeks after successful resuscitation (Fig. 3.40).

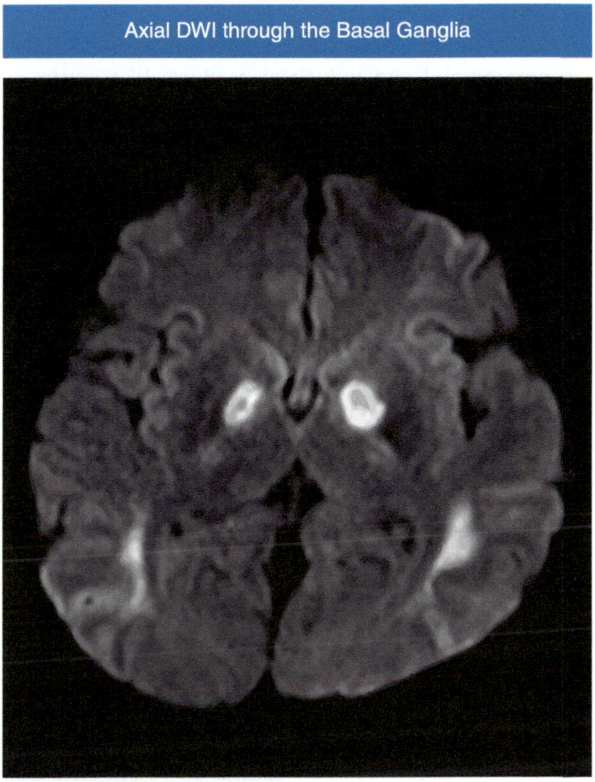

Fig. 3.40 Axial DWI through the brain of an individual who was found unconscious at their home with no additional clinical history. Symmetric diffusion restriction is seen within the bilateral globi pallidi. Additional posterior subcortical restricted diffusion also noted. These findings are suggestive of acute CO poisoning. However, cyanide poisoning and opiate toxicity can have overlapping imaging findings

- CT findings of CO poisoning:
  - **The globus pallidus is the most commonly affected structure in CO poisoning.**
    - CT scans typically reveal symmetric hypodensity of the globus pallidus.
  - There can be bilateral symmetric hypoattenuation in the caudate nucleus, putamen, and thalamus, but these structures are not as frequently damaged by CO poisoning.
- MRI findings of CO poisoning [126]:
  - The medial portions of the globi pallidi appear as bilateral areas of hypointensity on T1-weighted images and of hyperintensity on T2/FLAIR sequencing.
  - Other basal ganglia structures often exhibit asymmetric hyperintensities on T2-weighted and FLAIR images.
  - In the acute stage of CO poisoning, patchy or peripheral enhancement may be seen in necrotic areas.
  - T2-weighted images also show diffuse hyperintensities, especially in the centrum semiovale and periventricular white matter. Conversely, T1-weighted images show hypointensity in the same areas.

2. Ethylene Glycol

Ethylene glycol (EG) is a clear, odorless dihydroxyl alcohol used in many commercial and industrial products like antifreeze. In adults, it is sometimes ingested as a cheap substitute for ethanol or in cases of attempted suicide or murder. Children and pets may be exposed to accidental ingestion caused by decantation of ethylene glycol to unlabeled or open containers. This risk is further exacerbated by the fact that EG tastes sweet [127]. EG has a low toxicity itself but is broken down to four organic acids in vivo. These organic acids cause central nervous system depression, cardiopulmonary failure, and renal failure. Neurological symptoms consist of inebriation, ataxia, and, in severe cases, seizures and coma. After an initial latent phase, severe encephalopathy may develop, mimicking an acute basilar artery occlusion (Fig. 3.41).

- CT findings of ethylene glycol poisoning [128]:
  - The typical pattern seen on the initial CT scan is diffuse hypodensity of the bilateral basal ganglia and thalami extending into the brainstem and adjacent cerebral hemispheres with mild mass effect.
  - Hemorrhagic lesions of the bilateral globus pallidus or putamen have also been reported.

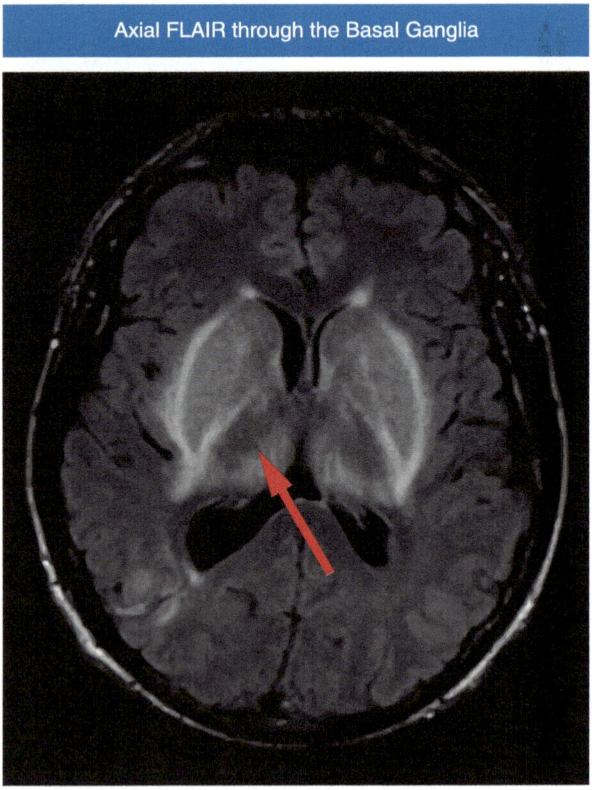

Fig. 3.41 An axial FLAIR image through the basal ganglia in a 20-something year old patient reveals bilateral basal ganglia signal hyperintensity. The arrow denotes additional prominent signal within the posterior limb of the internal capsule. This is called the **"lentiform fork" sign, and it can be seen in various causes of metabolic acidosis.** In this instance, it was accidental ingestion of ethylene glycol. No, I do not know how an adult can accidentally ingest antifreeze. This sign can also be seen in cases of uremic encephalopathy

- MRI findings of ethylene glycol poisoning [129, 130]:
  - Central basal portions of the brain exhibit bilaterally symmetric increased T2 signal, primarily in the thalami, basal ganglia, hippocampi, amygdala, dorsal midbrain, and pons.
  - This signal abnormality may also extend into adjacent portions of the cerebellum, brainstem, and insula, accompanied by mild mass effect.

3. Methanol Poisoning

Methanol is a colorless, flammable liquid commonly found in household solutions like aftershaves, perfumes, and cologne. Visual disturbance is the most striking acute symptom after ingestion. Other symptoms, such as drowsiness, seizures, coma, nausea, vomiting, respiratory arrest, and ultimately death, are usually delayed up to 24 hours because of delayed toxic metabolite production leading to metabolic acidosis. Methanol intoxication has a high mortality rate, and survivors experience long-term visual sequelae and severe brain damage. Word to the wise: *don't drink homemade hooch if it comes from a still or has undergone any other fortification process* [131].

- CT findings of methanol poisoning [132]:
  - The predominant pattern observed is bilateral and symmetric **putaminal** necrosis, either as a single lesion or in conjunction with other lesions. On CT scans, the putamina appear hypodense. Additionally, some patients had low attenuation lesions in the subcortical white matter of the insula.
  - **The timing of CT scans relative to methanol ingestion is crucial, as most abnormal findings were detected 24 hours post-ingestion.**
- MRI findings of methanol poisoning [133]:
  - The putamina are hyperintense on T2-weighted MR images and are commonly associated with T1 hyperintensity, reflecting hemorrhage.
  - In the acute phase, reduced diffusivity is present, and there may be peripheral contrast enhancement.
  - Look for preferential involvement:
  - Of the frontal white matter and insula.
  - While necrotic lesions in the basal ganglia are typically observed within the first 2–3 days of hospitalization, hemorrhagic components tend to appear later, sometimes 10–14 days post-admission.

4. Methotrexate Toxicity

Methotrexate (MTX) is used to treat inflammatory conditions like rheumatoid arthritis and some cancers such as acute lymphoblastic leukemia (ALL). The onset of neurotoxicity following induction ranges from 2 to 127 weeks. The risk factors for methotrexate related toxicity include pediatric age, high-doses, **intrathecal administration**, and concurrent cranial radiation. This is pertinent information because MTX is, with cytarabine, the most common intrathecal chemotherapeutic drug [134]. Clinical symptoms of MTX neurotoxicity are often associated with leukoencephalopathy, but it can also be a stroke mimic [135]. Germline polymorphisms, such as variants in GSTP1, 17 MTHFR, and SHMT1, may contribute to MTX-induced leukoencephalopathy and neurotoxicity. MTX-induced leukoencephalopathy has differing clinical presentations, including seizures, transient ischemic attacks, encephalopathy, movement disorders, and myelopathy.

- CT findings of MTX toxicity [136]:
  - In MTX-induced neurotoxicity, CT scans do not consistently capture abnormal findings.
  - CT scans are nonspecific for MTX-induced leukoencephalopathy; however, it usually presents with low attenuation in the white matter of both cerebral hemispheres.
  - Additional abnormal CT scan findings include calcifications in the basal ganglia and widening of subarachnoid spaces.
- MRI findings of MTX toxicity [137, 138]:
  - Leukoencephalopathy is the most common presentation with FLAIR signal hyperintensity in the subcortical and deep white matter. There is variable involvement of cortical U-fibers.
  - DWI shows restricted diffusion in anatomic brain regions associated with the symptoms, but T2-weighted and FLAIR images are less marked.
  - DWI findings normalize after recovery, whereas T2 and/or FLAIR imaging often reveal residual abnormalities (Fig. 3.42).

5. Radiation-Induced Leukoencephalopathy (RIL)

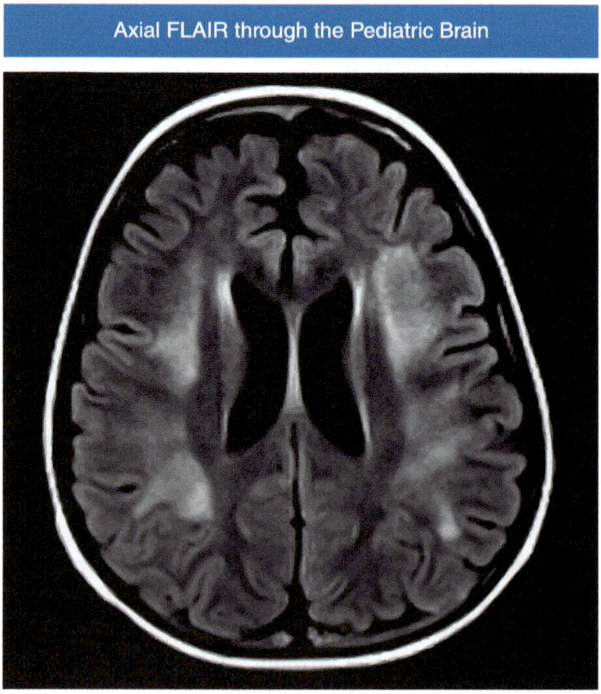

Fig. 3.42 Bilateral patchy signal hyperintensities are seen in this patient, which appear to spare the cortical U-fibers. These findings are quite nonspecific. The patient's clinical note revealed that they underwent chemotherapy for ALL 4 weeks prior. Methotrexate leukoencephalopathy should be a leading consideration

Radiation therapy is crucial in the treatment of head and neck tumors, either alone or in combination with chemotherapy and/or surgery. Some recognized risk factors of RIL are older age, diabetes, systemic hypertension, and combined radiation and chemotherapy regimens. Radiation-induced CNS complications may appear as an acute injury during treatment, an early or delayed complication within weeks to a few months after irradiation, or a late complication 6 months or longer after irradiation. The acute presentation of radiation is disseminated necrotizing leukoencephalopathy (DNL). The clinical symptoms are generally reversible in acute injury and do not predict later cognitive impairment. Most patients with subacute radiation-induced leukoencephalopathy are asymptomatic and do not present focal neurological deficits. RIL symptoms primarily present as cognitive dysfunction, particularly in areas of executive function, memory, and attention, impacting quality of life [139]. Clinical symptoms generally stabilize or evolve slowly in most cases of RIL.

- CT findings RIL:
  - On CT scans, radiation-induced leukoencephalopathy presents with rapidly progressive, symmetric hypodensities in the periventricular and deep white matter. It initially spares the corpus callosum and cortical U-fibers.
  - The lesions increase in size over time and combine to form a confluent pattern with smooth peripheral margins.
- MRI findings of RIL [140–142]:
  - On MRI, RIL is characterized by extensive, diffuse T2 FLAIR hyperintensities within the cerebral white matter with anterior predominance.
  - T2 signal hypointensities superimposed upon coalescent white matter FLAIR signal hyperintensity, which demonstrate concomitant contrast enhancement are suggestive of DNL.
  - The posterior fossa, internal capsule, basal ganglia, and subcortical U-fibers seem to be spared.
  - Long-term sequelae include induction of meningiomas, cavernous malformations, intracranial aneurysms, and steno-occlusive vasculopathy.

6. Opioids

   Worldwide, opioids are implicated in the greatest number of deaths associated with illicit drug use, and the United States leads in illicit opioid use and its associated morbidity and mortality. Nonfatal overdose estimates likely underestimate the true prevalence because survivors and bystanders usually do not seek medical care. An opioid overdose results in depressed mental status and/or unconsciousness, slow and shallow breathing, and constricted pupils. Other symptoms include apathy, memory impairment, and acute cerebellitis. Opioids are implicated in the disruption of myelin sheaths and/or axons, and opioid overdoses cause damage to brain areas sensitive to hypoxic ischemia, including the hippocampus and cerebellum. Necessary interventions usually include naloxone administration to reverse opioid agonist effects, respiratory support, and monitoring of mental status and breathing.

- CT findings of opioid toxicity:
  - CT scans of opioid-induced brain injury reveal bilateral hypodensities in the globi pallidi, indicating hypoxic-ischemic injury.
  - Delayed opioid-induced leukoencephalopathy presented with diffuse white matter hypodensities, sometimes appearing weeks after an overdose.
- MRI findings of opioid toxicity [143–145]:
  - Effects can be acute as in the setting of acute toxic leukoencephalopathy or delayed as in the case of delayed post-hypoxic leukoencephalopathy.
    - Both show confluent white matter FLAIR:
    - Signal hyperintensities.
      - The delayed variant shows additional cystic encephalomalacia.
    - Bilateral FLAIR signal hyperintensities in the hippocampus are possible.
  - Inhalation heroin toxicity, also known as "chasing the dragon," has some unique imaging findings.
    - FLAIR signal hyperintensity within the cerebellum.
    - Midbrain FLAIR signal hyperintensity that spares the substantia nigra and red nuclei.
    - FLAIR signal hyperintensities within the bilateral posterior limbs of the internal capsules (Figs. 3.43 and 3.44).

7. Ethanol

   To alcohol! The cause of—and solution to—all of life's problems. So sayeth Homer J. Simpson in an iconic, and one of my favorites, *Simpson's* episode. But on a serious note, ethanol is a poison. It's a poison that causes untold grief, violence, and legal woe around the world, but it is one that has been inextricably woven into the cultural and socioeconomic fabrics of multiple societies throughout human history. Therefore, it too is here because it is not going away. Going away is exactly what the brains of alcoholics do, especially the cerebellum. Florid atrophy of the brain in excess of the stated age is a common imaging finding in alcoholics.

- CT findings of alcohol abuse [146]:
  - Dilation of the ventricles
  - Atrophy of the cerebrum and cerebellum out of proportion to the patient's stated age.
- MRI findings of alcohol abuse [147]:
  - Wernicke Encephalopathy & Korsakoff syndrome (thiamine deficiency due to excess alcohol intake).
    - Periaqueductal gray and thalamic FLAIR signal hyperintensities
    - FLAIR signal hyperintensity and/or postcontrast T1 enhancement of the mammillary bodies
    - Thalamic and medial temporal lobe atrophy in advanced cases (Korsakoff Syndrome) (Figs. 3.45 and 3.46)

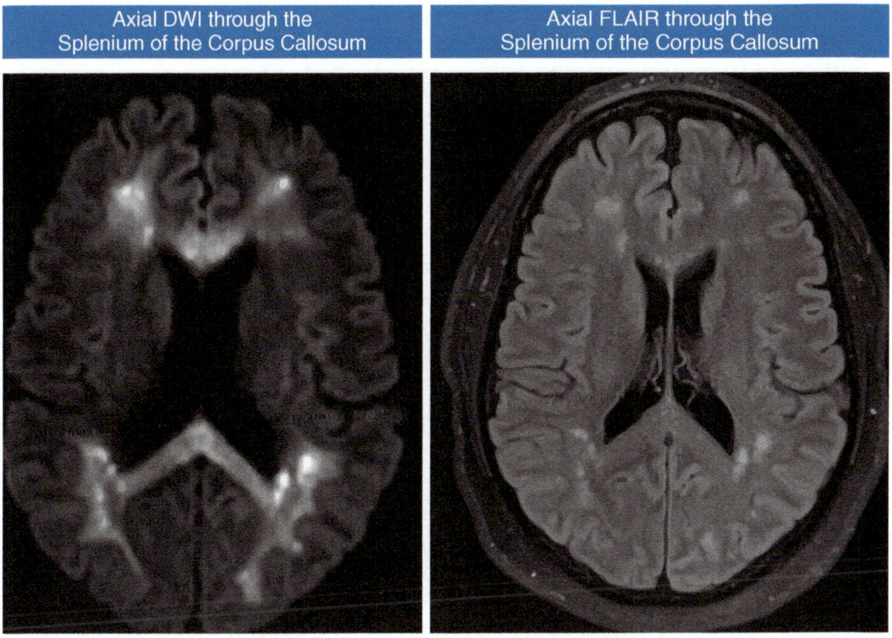

Fig. 3.43 MRI of the brain performed on a patient found down and unresponsive in their home. Notice the degree of restricted diffusion within the corpus callosum. The FLAIR sequencing is definitely not as impressive. The patient's toxicology results revealed fentanyl. They never regained consciousness

8. Hepatic Encephalopathy (HE)

   The liver is the great cleanser of the blood, receiving most of its supply from the portal venous circulation, which itself drains the guts. The remainder comes from the hepatic artery. Hepatic dysfunction can be the result of many different types of insult to the liver, including chronic alcohol ingestion, trauma, acute toxicity from acetaminophen ingestion, inborn errors of metabolism, and other toxins (just to name a few). With respect to the CNS, when the liver fails to do its job, HE ensues, which can result in neuropsychiatric dysfunction.

   - Motor and behavioral manifestations are common in this disease, which can be subdivided into three categories based on duration and characteristics: episodic, chronic, and minimal [148].
     – The pathophysiology of HE is complex and beyond the scope of this text, but suffice it to say that glutamate and GABA neurotransmitter imbalance is a primary contributor.
     – Excess ammonia, which is found in excess in unfiltered portal blood, is implicated.
     – Excess manganese deposition in the brain is thought to be neurotoxic.
   - The imaging characteristics of HE are interesting, especially on MRI [149].

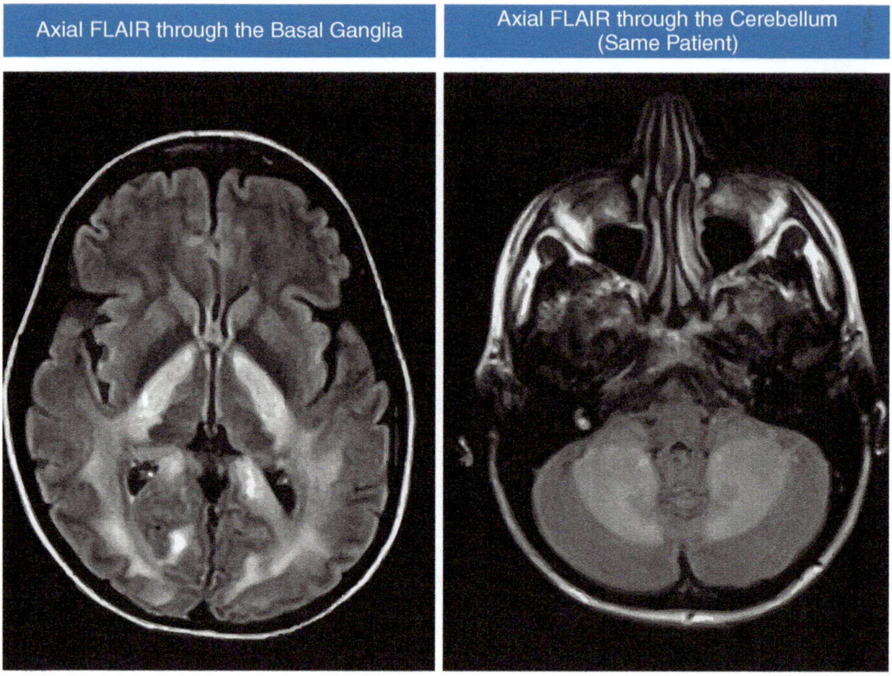

Fig. 3.44 This patient presented in an obtunded state. Initial CT revealed symmetric posterior hypoattenuation. Subsequent MR imaging was obtained. The axial FLAIR image on your left reveals symmetric FLAIR signal hyperintensities in the posterior limbs of both internal capsules. The axial FLAIR image on your right through the cerebellum shows symmetric hyperintensities in both cerebellar hemispheres. These findings are strongly suggestive of inhalational opioid toxicity also known as chasing the dragon. ***The dragon will work to elevate you but the consequences are dire***

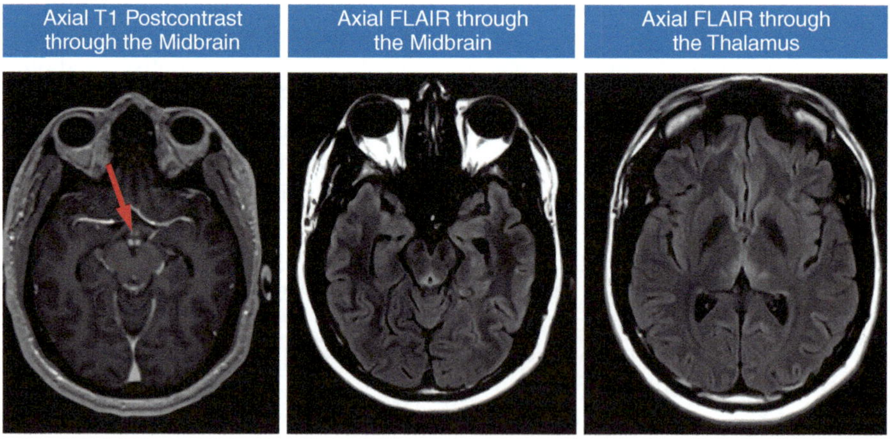

Fig. 3.45 Left postcontrast T1 image through the midbrain reveals uniform bilateral enhancement of the mammillary bodies. The middle axial FLAIR image shows signal hyperintensity surrounding the ventricle. The axial FLAIR image on your right shows signal hyperintensity of the bilateral periventricular thalami. Remember clinical context. These findings are similar to NMOSD, with the exception of the mammillary body enhancement. This was a case of Wernicke's encephalopathy

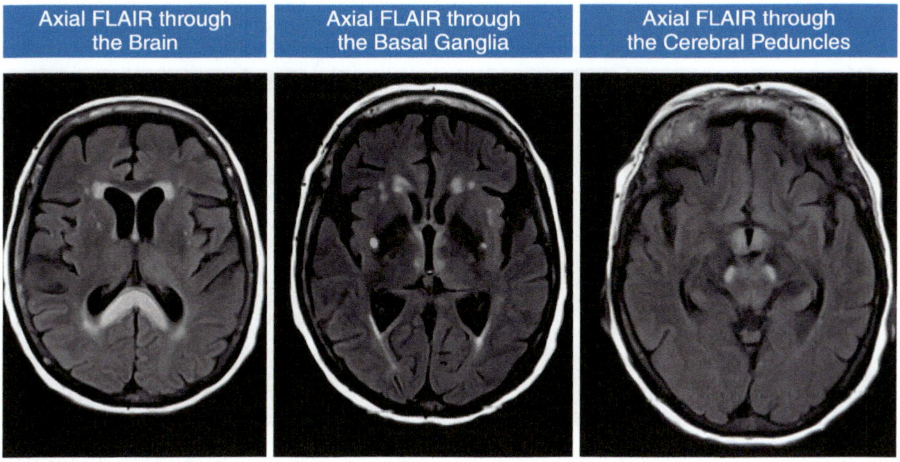

Fig. 3.46 MRI of the brain of an intractable alcoholic. The image on your left reveals pronounced FLAIR signal hyperintensity within the splenium of the corpus callosum. The middle image shows FLAIR signal hyperintensity flanking the third ventricular parenchyma. The right image shows FLAIR signal hyperintensity within the mammillary bodies and both medial lemnisci. These findings suggest the patient suffers from both Wernicke encephalopathy **and** Marchiafava-Bignami disease

- Intrinsic T1 signal hyperintensity can be seen in the basal ganglia, specifically the globus pallidus, as well as the substantia nigra due to manganese deposition, which is paramagnetic.
- Acute HE can show cortical FLAIR signal hyperintensity and restricted diffusion, specifically involving the cingulate gyrus and insular cortices (Fig. 3.47).

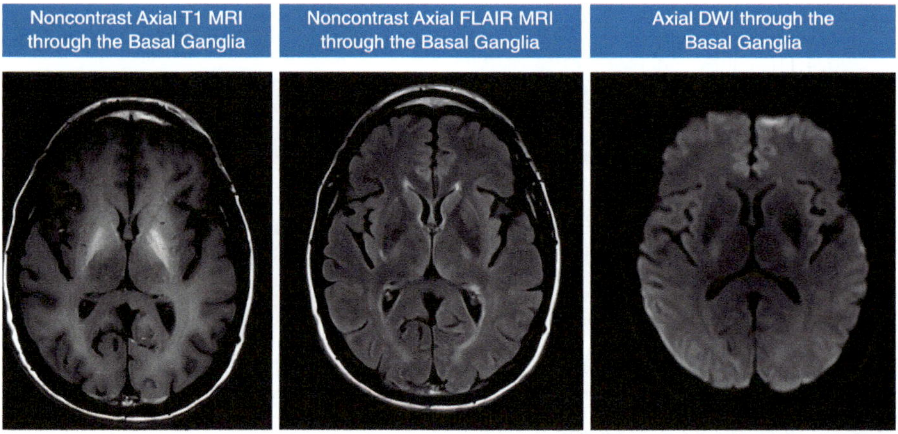

Fig. 3.47 An encephalopathic patient undergoes MRI and you receive the following sequences. By now, you're probably partial to FLAIR sequences for most of your diagnostic imaging. Not so in this case. The precontrast T1 signal hyperintensity in the basal ganglia should invoke a differential diagnosis. Hepatic encephalopathy is one. Some other examples include total parenteral nutrition, and toxic metallic exposure in welders. Other causes of T1 shortening in the basal ganglia are NF1 and Japanese encephalitis

References

1. Wang H, Zhao S, Wang S, Zheng Y, Wang S, Chen H, et al. Global magnitude of encephalitis burden and its evolving pattern over the past 30 years. J Infect. 2022;84(6) https://doi.org/10.1016/j.jinf.2022.04.026.
2. Rouse BT, Sehrawat S, Rouse BT, Sehrawat S. Immunity and immunopathology to viruses: what decides the outcome? Nat Rev Immunol. 2010;10(7) https://doi.org/10.1038/nri2802.
3. Riccardi N, Antonello RM, Luzzati R, Zajkowska J, Bella SD, Giacobbe DR. Tick-borne encephalitis in Europe: a brief update on epidemiology, diagnosis, prevention, and treatment. Eur J Intern Med. 2019:62. https://doi.org/10.1016/j.ejim.2019.01.004.
4. Fayisa ST, Pokharel N, Solomon A, Negash L, Biltibo LA, Bedasso SH, et al. Navigating diagnostic challenges: imaging strategies for herpes simplex encephalitis in resource limited settings: a case report. Radiol Case Rep. 2024;19(6) https://doi.org/10.1016/j.radcr.2024.02.060.
5. Zimmerman RD, Russell EJ, Leeds NE. D Kaufman CT in the early diagnosis of herpes simplex encephalitis. AJR Am J Roentgenol. 1980;134(1) https://doi.org/10.2214/ajr.134.1.61.
6. Tunkel AR, Glaser CA, Bloch KC, Sejvar JJ, Marra CM, Roos KL, et al. The Management of Encephalitis: clinical practice guidelines by the Infectious Diseases Society of America. Clin Infect Dis. 2008;47(3) https://doi.org/10.1086/589747.
7. Koksel Y, McKinney AM. Potentially reversible and recognizable acute Encephalopathic syndromes: disease categorization and MRI appearances. Am J Neuroradiol. 2020; https://doi.org/10.3174/ajnr.A6634.
8. Sarton B, Jaquet P, Belkacemi D, et al. MRI changes and functional outcome among adults with severe herpes simplex encephalitis. JAMA Netw Open. 2021;4(7) https://doi.org/10.1001/jamanetworkopen.2021.14328.
9. Chow FC, Glaser CA, Sheriff H, Xia D, Messenger S, Whitley R, et al. Use of clinical and neuroimaging characteristics to distinguish temporal lobe herpes simplex encephalitis from its mimics. Clin Infect Dis. 2015;60(9) https://doi.org/10.1093/cid/civ051.

10. Webb G, Leong MYM, Bishop E, Sehu M. Diagnostic dilemmas: a review of reported cases of human herpesvirus 6 encephalitis in immunocompetent adults. Open Forum Infect Dis. 2024;11(9) https://doi.org/10.1093/ofid/ofae501.
11. Noguchi T, Yoshiura T, Hiwatashi A, Togao O, Yamashita K, Nagao E, et al. CT and MRI findings of human herpesvirus 6–associated encephalopathy: comparison with findings of herpes simplex virus encephalitis. Am J Roentgenol. 2012;194(3) https://doi.org/10.2214/AJR.09.2548.
12. Berzero G, Campanini G, Vegezzi E, Paoletti M, Pichiecchio A, Simoncelli AM, Colombo AA. Human herpesvirus 6 encephalitis in immunocompetent and immunocompromised hosts. Neurol Neuroimmunol Neuroinflamm. 2021;8(2) https://doi.org/10.1212/NXI.0000000000000942.
13. Hills SL, Broussard KR, Broyhill JC, Shastry LG, Cossaboom CM, White JL, et al. Tick-borne encephalitis among US travellers, 2010–20. J Travel Med. 2022;29(2) https://doi.org/10.1093/jtm/taab167.
14. Pustijanac E, Buršić M, Talapko J, Škrlec I, Meštrović T, Lišnjić D, et al. Tick-borne encephalitis virus: a comprehensive review of transmission, pathogenesis, epidemiology, clinical manifestations, diagnosis, and prevention. Microorganisms. 2023;11(7) https://doi.org/10.3390/microorganisms11071634.
15. Horger M, Beck R, Fenchel M, Ernemann U, Nägele T, Brodoefel H, et al. Imaging findings in tick-borne encephalitis with differential diagnostic considerations. Am J Roentgenol. 2012;199(2) https://doi.org/10.2214/AJR.11.7911.
16. Zawadzki R, Garkowski A, Kubas B, Zajkowska J, Hładuński M, Jurgilewicz D, et al. Evaluation of imaging methods in tick-borne encephalitis. Pol J Radiol. 2017:82. https://doi.org/10.12659/PJR.903940.
17. Wang H, Liang G. Epidemiology of Japanese encephalitis: past, present, and future prospects. Ther Clin Risk Manag. 2015:11. https://doi.org/10.2147/TCRM.S51168.
18. Pichl T, Wedderburn CJ, Hoskote C, Turtle L, Bharucha T. A systematic review of brain imaging findings in neurological infection with Japanese encephalitis virus compared with dengue virus. Int J Infect Dis. 2022:119. https://doi.org/10.1016/j.ijid.2022.03.010.
19. Phukan P, Sarma K, Sharma BK, Boruah DK, Gogoi BB, Chuunthang D. MRI Spectrum of Japanese encephalitis in Northeast India: a cross-sectional study. Journal of neurosciences in rural. Practice. 2021;12(2) https://doi.org/10.1055/s-0041-1722820.
20. Lucas S. Historical and current issues in HIV encephalitis, and the role of neuropathology in HIV disease: a pathological perspective. J Neurol. 2022:270(3). https://doi.org/10.1007/s00415-022-11503-2.
21. Given CA. Neuroimaging of the HIV/AIDS patient. HIV/AIDS and the nervous. System. 2007:85. https://doi.org/10.1016/S0072-9752(07)85016-0.
22. HIV Encephalopathy (Chapter 25) – Brain Imaging with MRI and CT; 2012. doi: https://doi.org/10.1017/CBO9781139030854.026.
23. Sakai M, Higashi M, Fujiwara T, Uehira T, Shirasaka T, Nakanishi K, et al. MRI imaging features of HIV-related central nervous system diseases: diagnosis by pattern recognition in daily practice. Jpn J Radiol. 2021;39(11) https://doi.org/10.1007/s11604-021-01150-4.
24. Baldassari LE, Wattjes MP, Cortese ICM, Gass A, Metz I, Yousry T, et al. The neuroradiology of progressive multifocal leukoencephalopathy: a clinical trial perspective. Brain. 2021;145(2) https://doi.org/10.1093/brain/awab419.
25. Paul S, Khanam M, Khan MAS, Khan S, Ahmed MZ, Moontaha H, et al. Rabies encephalitis following post exposure prophylaxis (PEP) is becoming an evolving problem in Bangladesh. Microbe. 2024:3. https://doi.org/10.1016/j.microb.2024.100094.
26. Jackson A. Rabies encephalitis. Int J Infect Dis. 2016:45. https://doi.org/10.1016/j.ijid.2016.02.089.
27. Jassi P, Attri A, Dhawan R, Kakkar C, Saggar K. MR imaging in rabies encephalitis: a rare entity. Annals of Indian academy of. Neurology. 2016;19(1) https://doi.org/10.4103/0972-2327.167712.

28. Jain H, Deshpande A, Favaz AM, Rajagopal KV. MRI in rabies encephalitis. Case Reports. 2013:30. https://doi.org/10.1136/bcr-2013-201825.
29. Gutierrez J, Issacson RS. Subacute sclerosing panencephalitis: an update. Dev Med Child Neurol. 2010;52(10) https://doi.org/10.1111/j.1469-8749.2010.03717.x.
30. Keerthiraj DB, Pandey S, Kumar Garg R, Singh Malhotra H, Verma R, Kumar Sharma P, et al. Neuroimaging abnormalities in patients with subacute Sclerosing Panencephalitis. Clin Neuroradiol. 2024;34(3) https://doi.org/10.1007/s00062-024-01396-1.
31. Barthwal S, Kaur N, Kaur R, Zaidi R, Randev S. Magnetic resonance imaging in subacute Sclerosing Panencephalitis: two case reports and review of literature. Cureus. 2021;13(10) https://doi.org/10.7759/cureus.19161.
32. Phang DSK, Ettikan JV, Aziz HA, Vendargon FM, Teo KSS. A rare complication of herpes zoster Ophthalmicus (HZO). Cureus. 2023;15(3) https://doi.org/10.7759/cureus.35693.
33. Bae DW, An JY. Herpes zoster Ophthalmicus presenting as acute orbital inflammation preceding the vesicular rash. J Neurosonol Neuroimag. 2020;12(2) https://doi.org/10.31728/jnn.2020.00090.
34. Yong H, Wallace C, Kapadia R. "Trigeminal tract sign" in patients with herpes zoster Ophthalmicus: a case series of a novel imaging finding. Neurology. 2022;99(23_Supplement_2) https://doi.org/10.1212/01.wnl.0000903564.62886.c1.
35. Soares BP, Provenzale JM, Soares BP, Provenzale JM. Imaging of herpesvirus infections of the CNS. Am J Roentgenol. 2015;206(1) https://doi.org/10.2214/AJR.15.15314.
36. Nago B, Liu JE. An unusual case of varicella zoster encephalitis mimicking a glioblastoma on magnetic resonance imaging and magnetic resonance spectroscopy. IDCases. 2025:39. https://doi.org/10.1016/j.idcr.2024.e02124.
37. Amlie-Lefond C, Gilden D. Varicella zoster virus: a not uncommon cause of stroke in children and adults. J Stroke Cerebrovasc Dis. 2016;25(7) https://doi.org/10.1016/j.jstrokecerebrovasdis.2016.03.052.
38. Goswami Y, Gaurkar SS. Ramsay hunt syndrome: An introduction, signs and symptoms, and treatment. Cureus. 2023;15(1) https://doi.org/10.7759/cureus.33688.
39. Ungureanu A, Jvd M, Bicvic A, Abbuehl L, Chiffi G, Jaques L, et al. Meningitis, meningoencephalitis and encephalitis in Bern: an observational study of 258 patients. BMC Neurol. 2021;21(1) https://doi.org/10.1186/s12883-021-02502-3.
40. Runde T, Anjum F, Hafner JW. Bacterial meningitis. In:Treasure Island, FL: StatPearls Publishing. August 8, 2023. https://www.ncbi.nlm.nih.gov/books/NBK470351/. Accessed 7 Feb 2025.
41. Salazar L, Hasbun R. Cranial imaging before lumbar puncture in adults with community-acquired meningitis: clinical utility and adherence to the Infectious Diseases Society of America guidelines. Clin Infect Dis. 2017;64(12) https://doi.org/10.1093/cid/cix240.
42. Carter E, McGill F. The management of acute meningitis: an update. Clin Med. 2022;22(5) https://doi.org/10.7861/clinmed.2022-cme-meningitis.
43. Mohan S, Jain KK, Arabi M, Shah GV. Imaging of meningitis and ventriculitis. Neuroimaging Clin N Am. 2012;22(4) https://doi.org/10.1016/j.nic.2012.04.003.
44. Prodi E, Danieli L, Manno C, Pagnamenta A, Pravatà E, Roccatagliata L, et al. Stroke mimics in the acute setting: role of multimodal CT protocol. Am J Neuroradiol. 2022;43(2) https://doi.org/10.3174/ajnr.A7379.
45. Beuker C, Werring N, Bonberg N, Strecker JK, Schmidt-Pogoda A, Schwindt W, Stracke P, et al. Stroke in patients with bacterial meningitis: a cohort study and meta-analysis. Ann Neurol. 2023;93(6) https://doi.org/10.1002/ana.26618.
46. Brouwer MC, Coutinho JM, van de Beek D. Clinical characteristics and outcome of brain abscess. Neurology. 2014;82(9) https://doi.org/10.1212/WNL.0000000000000172.
47. Rath TJ, Hughes M, Arabi M, Shah GV. Imaging of cerebritis, encephalitis, and brain abscess. Neuroimaging Clin N Am. 2012;22(4) https://doi.org/10.1016/j.nic.2012.04.002.
48. Feraco P, Donner D, Gagliardo C, Leonardi I, Piccinini S, Del Poggio A, Franciosi R, et al. Cerebral abscesses imaging: a practical approach. J Popul Ther Clin Pharmacol. 2020:27(3). https://doi.org/10.15586/jptcp.v27i3.688.

49. Nakayama M, Naganawa S, Ouyang M, Jones KA, Kim J, Capizzano AA, et al. A review of clinical and imaging findings in Tumefactive demyelination. Am J Roentgenol. 2021;217(1) https://doi.org/10.2214/AJR.20.23226.
50. Honda H, Warren DK. Central nervous system infections: meningitis and brain abscess. Infect Dis Clin N Am. 2009;23(3) https://doi.org/10.1016/j.idc.2009.04.009.
51. Lee WJ, Tsai MH, Hsu JF, Chu SM, Chen CC, Yang PH, Huang HR, Chi MC, Lee CW. The epidemiology, management and therapeutic outcomes of subdural empyema in neonates with acute bacterial meningitis. Antibiotics (Basel, Switzerland). 2024;13(4) https://doi.org/10.3390/antibiotics13040377.
52. Watson C, Post N, Camacho A. Subdural empyema mimicking subacute subdural hematoma on CT imaging. Interdisciplinary. Neurosurgery. 2018:13. https://doi.org/10.1016/j.inat.2018.04.002.
53. Wong AM, Zimmerman RA, Simon EM, Pollock AN, Bilaniuk LT. Diffusion-weighted MR imaging of subdural empyemas in children. AJNR Am J Neuroradiol. 2004;25(6):1016–21.
54. Rose DZ, Parra-Herran C, Petito CK, Post MJD. Restricted diffusion of pus in the subarachnoid space: MRSA Meningo-Vasculitis and progressive brainstem ischemic strokes – a case report. Case Rep Neurol. 2010;2(2) https://doi.org/10.1159/000319691.
55. Luque-Paz D, Revest M, Eugène F, Boukthir S, Dejoies L, Tattevin P, et al. Ventriculitis: a severe complication of central nervous system infections. Open forum. Infect Dis. 2021;8(6) https://doi.org/10.1093/ofid/ofab216.
56. Ramanan M, Shorr A, Lipman J. Ventriculitis: infection or inflammation. Antibiotics. 2021;10(10) https://doi.org/10.3390/antibiotics10101246.
57. Fujikawa A, Tsuchiya K, Honya K, Nitatori T, Fujikawa A, Tsuchiya K, et al. Comparison of MRI sequences to detect Ventriculitis. Am J Roentgenol. 2012;187(4) https://doi.org/10.2214/AJR.04.1923.
58. Kuo I, Long T, Nguyen N, Chaudry B, Karp M, Sanossian N. Ruptured intracranial mycotic aneurysm in infective endocarditis: a natural history. Case Rep Med. 2010;2010(1) https://doi.org/10.1155/2010/168408.
59. Wang TKM, Griffin B, Cremer P, Shrestha N, Gordon S, Pettersson G, Desai M, et al. Diagnostic utility of CT and MRI for mycotic aneurysms: a meta-analysis. AJR Am J Roentgenol. 2020;215(5) https://doi.org/10.2214/AJR.19.22722.
60. Shen G, Shen X, Pu W, Zhang G, Lerner A, Gao B. Imaging of cerebrovascular complications of infection. Quant Imaging Med Surg. 2018;8(10) https://doi.org/10.21037/qims.2018.11.08.
61. John S, Walsh KM, Hui FK, Sundararajan S, Silverman S, Bain M. Dynamic angiographic nature of cerebral mycotic aneurysms in patients with infective endocarditis. Stroke. 2016;47(1) https://doi.org/10.1161/STROKEAHA.115.011198.
62. Donoghue HD, Lee OY-C, Minnikin DE, Besra GS, Taylor JH, Spigelman M. Tuberculosis in Dr Granville's mummy: a molecular re-examination of the earliest known Egyptian mummy to be scientifically examined and given a medical diagnosis. Proc R Soc B Biol Sci. 2009;277(1678) https://doi.org/10.1098/rspb.2009.1484.
63. Seddon JA, Tugume L, Solomons R, Prasad K, Bahr NC, Consortium TMIR. The current global situation for tuberculous meningitis: epidemiology, diagnostics, treatment and outcomes. Wellcome Open Research. 2019:4. https://doi.org/10.12688/wellcomeopenres.15535.1.
64. Dian S, Hermawan R, et al. Brain MRI findings in relation to clinical characteristics and outcome of tuberculous meningitis. PLoS One. 2020;15(11) https://doi.org/10.1371/journal.pone.0241974.
65. Choudhary N, Vyas S, Modi M, Raj S, Kumar A, Sankhyan N, et al. MR vessel wall imaging in tubercular meningitis. Neuroradiology. 2021;63(10) https://doi.org/10.1007/s00234-021-02678-y.
66. Huang Y, Zhang Y, Yang S, Lu H, Yu H, Wang X, et al. Epidemiology of cryptococcal meningitis and fluconazole heteroresistance in Cryptococcus neoformans isolates from a teaching hospital in southwestern China. Microbiol Spectr. 2024;12(8) https://doi.org/10.1128/spectrum.00725-24.

67. Monsalve GP, Forero N. Cryptococcal meningitis and spatial distribution of pigeons (Columba livia) in Colombia (P4-8.004). Neurology. 2023;100(17_supplement_2):4760.
68. Anjum SH, Bennett JE, Dean O, Marr KA, Hammoud DA, Williamson PR, et al. Neuroimaging of Cryptococcal meningitis in patients without human immunodeficiency virus: data from a multi-center cohort study. J Fungi. 2023;9(5):594. https://doi.org/10.3390/jof9050594.
69. Trinh K, Le D, Kuo A. Teaching NeuroImage: Cryptococcal meningoencephalitis with Cryptococcoma and gelatinous pseudocysts. Neurology. 2023;101(7) https://doi.org/10.1212/WNL.0000000000207359.
70. Wang H, Zhang Q, Shi Y, Li R, Wang S, Sun J, et al. The imaging diagnostic criteria of AIDS-related cerebral toxoplasmosis in China. Radiology of. Infect Dis. 2020;7(3) https://doi.org/10.1016/j.jrid.2020.08.003.
71. Marcus C, Feizi P, Hogg J, Summerfield H, Castellani R, Sriwastava S, et al. Imaging in differentiating cerebral toxoplasmosis and primary CNS lymphoma with special focus on FDG PET/CT. Am J Roentgenol. 2020; https://doi.org/10.2214/AJR.19.22629.
72. Pai V, Sansi R, Kharche R, Bandili SC, Pai B, Pai V, et al. Rhino-orbito-cerebral Mucormycosis: pictorial review. Insights Imaging. 2021;12(1) https://doi.org/10.1186/s13244-021-01109-z.
73. Ibrahim AS, Spellberg B, Walsh TJ, Kontoyiannis DP. Pathogenesis of Mucormycosis. Clin Infect Dis. 2012;54(Suppl_1) https://doi.org/10.1093/cid/cir865.
74. Vasudevan B, Hazra N, Shijith K, Neema S, Vendhan S. Mucormycosis: the scathing invader. Indian J Dermatol. 2021;66(4) https://doi.org/10.4103/ijd.ijd_477_21.
75. Balai E, Mummadi S, Jolly K, Darr A, Aldeerawi H. Rhinocerebral Mucormycosis: a ten-year single Centre case series. Cureus. 2020;12(11) https://doi.org/10.7759/cureus.11776.
76. Kaushal D, Rajan N, Soni K, Sharma A, Choudhury B, Yadav T, et al. Reducing mortality in mucormycosis of the head and neck in diabetic patients: a CARE case series. Eur Ann Otorhinolaryngol Head Neck Dis. 2022;139(3) https://doi.org/10.1016/j.anorl.2021.06.015.
77. Kondapavuluri SK, et al. Spectrum of MR imaging findings of sinonasal mucormycosis in post COVID-19 patients. Br J Radiol. 2021;94(1127) https://doi.org/10.1259/bjr.20210648.
78. Kovacs GG, Budka H. Molecular pathology of human prion diseases. Int J Mol Sci. 2009;10(3) https://doi.org/10.3390/ijms10030976.
79. Alpers MP. The epidemiology of kuru: monitoring the epidemic from its peak to its end. Philos Trans R Soc B Biol Sci. 2008;363(1510) https://doi.org/10.1098/rstb.2008.0071.
80. Bizzi A, Pascuzzo R, Blevins J, Grisoli M, Lodi R, Moscatelli MEM, et al. Evaluation of a new criterion for detecting prion disease with diffusion MRI. JAMA Neurol. 2020;77(9) https://doi.org/10.1001/jamaneurol.2020.1319.
81. Jesuthasan A, Sequeira D, Hyare H, Odd H, Rudge P, Mok TH, et al. Assessing initial MRI reports for suspected CJD patients. J Neurol. 2022;269(8) https://doi.org/10.1007/s00415-022-11087-x.
82. Perani D, Cortelli P. Neuroimaging and fatal familial insomnia. In: Neuroimaging of sleep and sleep disorders. Cambridge: Cambridge University Press; 2013. p. 375–83.
83. Walton C, King R, Rechtman L, Kaye W, Leray E, Marrie RA, et al. Rising prevalence of multiple sclerosis worldwide: insights from the atlas of MS, 3rd edition. Multiple Sclerosis (Basingstoke. 2020 ;26(14). doi: https://doi.org/10.1177/1352458520970841.
84. Simpson S, Wang W, Otahal P, Blizzard L, van der Mei IAF, Taylor BV. Latitude continues to be significantly associated with the prevalence of multiple sclerosis: an updated meta-analysis. J Neurol Neurosurg Psychiatry. 2019;90(11) https://doi.org/10.1136/jnnp-2018-320189.
85. Filippi M, Rocca MA. Acute disseminated encephalomyelitis. White matter diseases. 2020. doi: https://doi.org/10.1007/978-3-030-38621-4_5.
86. Lu Z, Zhang B, Qiu W, Kang Z, Shen L, Long Y, et al. Comparative brain stem lesions on MRI of acute disseminated encephalomyelitis, Neuromyelitis Optica, and multiple sclerosis. PLoS One. 2011;6(8) https://doi.org/10.1371/journal.pone.0022766.
87. Wang CX, Wang CX. Assessment and Management of Acute Disseminated Encephalomyelitis (ADEM) in the pediatric patient. Pediatr Drugs. 2021;23(3) https://doi.org/10.1007/s40272-021-00441-7.

88. Alsaid HM, Atawneh MA, Abukhalaf S, Daoud A, Hamadah A, Gharaibeh K. Acute hemorrhagic Leukoencephalitis – a rare but fatal form of acute disseminated encephalomyelitis – complicated by brain herniation: a case report and literature review. Am J Case Rep. 2022:23. https://doi.org/10.12659/AJCR.935636.
89. Mir F, Dar W, Yaqoob A, Wani M, Asmi R, Wani M, et al. Seroprevalence of campylobacter jejuni infection in common subtypes of Guillain–Barre syndrome in Kashmiri population. Egypt J Neurol Psychiatry Neurosurg. 2023;59(1) https://doi.org/10.1186/s41983-023-00764-x.
90. Bragazzi NL, Kolahi A-A, Nejadghaderi SA, Lochner P, Brigo F, Naldi A, et al. Global, regional, and national burden of Guillain–Barré syndrome and its underlying causes from 1990 to 2019. J Neuroinflammation. 2021;18(1) https://doi.org/10.1186/s12974-021-02319-4.
91. Alkan O, Yildirim T, Tokmak N, Tan M. Spinal MRI findings of Guillain-Barré syndrome. J Radiol Case Rep. 2009;3(3) https://doi.org/10.3941/jrcr.v3i3.153.
92. Grimm A, Oertl H, Auffenberg E, Schubert V, Ruschil C, Axer H, et al. Differentiation between Guillain–Barré syndrome and acute-onset chronic inflammatory demyelinating Polyradiculoneuritis – a prospective follow-up study using ultrasound and neurophysiological measurements. Neurotherapeutics. 2019;16(3) https://doi.org/10.1007/s13311-019-00716-5.
93. Banwell B, Bennett JL, Marignier R, Kim HJ, Brilot F, Flanagan EP, et al. Diagnosis of myelin oligodendrocyte glycoprotein antibody-associated disease: international MOGAD panel proposed criteria. Lancet Neurol. 2023;22(3) https://doi.org/10.1016/S1474-4422(22)00431-8.
94. Reindl M, Waters P, Reindl M, Waters P. Myelin oligodendrocyte glycoprotein antibodies in neurological disease. Nat Rev Neurol. 2018;15(2) https://doi.org/10.1038/s41582-018-0112-x.
95. Hor JY, Fujihara K. Epidemiology of myelin oligodendrocyte glycoprotein antibody-associated disease: a review of prevalence and incidence worldwide. Front Neurol. 2023:14. https://doi.org/10.3389/fneur.2023.1260358.
96. Baumann M, Grams A, Djurdjevic T, Wendel EM, Lechner C, Behring B, Blaschek A, et al. MRI of the first event in pediatric acquired demyelinating syndromes with antibodies to myelin oligodendrocyte glycoprotein. J Neurol. 2018;265(4) https://doi.org/10.1007/s00415-018-8781-3.
97. Bartels F, Lu A, Oertel FC, Finke C, Paul F, Chien C. Clinical and neuroimaging findings in MOGAD–MRI and OCT. Clin Exp Immunol. 2021;206(3) https://doi.org/10.1111/cei.13641.
98. Guo Y, Lennon VA, Parisi JE, Popescu B, Vasquez C, Pittock SJ, et al. Spectrum of sublytic astrocytopathy in neuromyelitis optica. Brain. 2022;145(4) https://doi.org/10.1093/brain/awab394.
99. Hor JY, Asgari N, Nakashima I, Broadley SA, Leite MI, Kissani N, et al. Epidemiology of Neuromyelitis Optica Spectrum disorder and its prevalence and incidence worldwide. Front Neurol. 2020:11. https://doi.org/10.3389/fneur.2020.00501.
100. Kim HJ, Paul F, Lana-Peixoto MA, Tenembaum S, Asgari N, Palace J, et al. MRI characteristics of neuromyelitis optica spectrum disorder: An international update. Neurology. 2015;84(11) https://doi.org/10.1212/WNL.0000000000001367.
101. Carandini T, Sacchi L, Bovis F, Azzimonti M, Bozzali M, Galimberti D, et al. Distinct patterns of MRI lesions in MOG antibody disease and AQP4 NMOSD: a systematic review and meta-analysis. Mult Scler Relat Disord. 2021:54. https://doi.org/10.1016/j.msard.2021.103118.
102. Triplett JD, Kermode AG. TNF-alpha inhibitor therapy associated demyelination. J Neurol Neurosurg Psychiatry. 2017;88(5) https://doi.org/10.1136/jnnp-2017-316074.54.
103. Taylor TR, Galloway J, Davies R, Hyrich K, Dobson R. Demyelinating events following initiation of anti-TNFα therapy in the British Society for Rheumatology biologics registry in rheumatoid arthritis. Neurol Neuroimmunol Neuroinflamm. 2021;8(3):e992.
104. Kucharz EJ, Kotulska-Kucharz A. Tumor necrosis factor alpha inhibitors and demyelinating disease: what is behind it? Reumatologia. 2021;59(2) https://doi.org/10.5114/reum.2021.105438.

105. Babin M, Golse M, Khaterchi M, Bapst B, Ancelet C, Nasser G, et al. Perivascular enhancement pattern: identification, diagnostic spectrum and practical approach – a pictorial review. J Neuroradiol. 2025;52(2) https://doi.org/10.1016/j.neurad.2025.101242.
106. Greco A, Virgilio AD, Gallo A, Fusconi M, Turchetta R, Tombolini M, et al. Susac's syndrome – pathogenesis, clinical variants and treatment approaches. Autoimmun Rev. 2014;13(8) https://doi.org/10.1016/j.autrev.2014.04.004.
107. David C, Sacré K, Henri-Feugeas M-C, Klein I, Doan S, Cohen FA, et al. Susac syndrome: a scoping review. Autoimmun Rev. 2022;21(6) https://doi.org/10.1016/j.autrev.2022.103097.
108. Cviková M, Štefela J, Všianský V, Dufek M, Doležalová I, Vinklárek J, et al. Frontiers | case report: Susac syndrome – two ends of the spectrum, single center case reports and review of the literature. Front Neurol. 2024:15. https://doi.org/10.3389/fneur.2024.1339438.
109. Yahyavi-Firouz-Abadi N, Kiczek M, Zeiler SR, Wasserman BA. Imaging features of Susac syndrome on high-resolution intracranial Vessel Wall MRI. Neurol Neuroimmunol Neuroinflamm. 2021;8(4) https://doi.org/10.1212/NXI.0000000000000994.
110. Bradshaw MJ, Pawate S, Koth LL, Cho TA, Gelfand JM. Neurosarcoidosis: pathophysiology, diagnosis, and treatment. Neurol Neuroimmunol Neuroinflamm. 2021;8(6):e1084.
111. Rossides M, Darlington P, Kullberg S, Arkema EV. Sarcoidosis: epidemiology and clinical insights. J Intern Med. 2023;293(6):668–80.
112. Sambon P, Sellimi A, Kozyreff A, Gheysens O, Pothen L, Yildiz H, et al. Frontiers | epidemiology, clinical presentation, treatment, and outcome of neurosarcoidosis: a mono-centric retrospective study and literature review. Front Neurol. 2022:13. https://doi.org/10.3389/fneur.2022.970168.
113. Ginat DT, Dhillon G, Almast J. Magnetic resonance imaging of Neurosarcoidosis. J Clin Imaging Sci. 2011:1. https://doi.org/10.4103/2156-7514.76693.
114. Cicilet S, Reddy S, Kancharla M. Insights into neurosarcoidosis: an imaging perspective. Pol J Radiol. 2023;88(1) https://doi.org/10.5114/pjr.2023.134021.
115. Nair JR, Moots RJ. Behcet's disease. Clin Med. 2017;17:1. https://doi.org/10.7861/clinmedicine.17-1-71.
116. Pretorius JJ, Hiesgen J, Myburgh M, Suleman FE. Neuro-Behçet – clinical and radiological findings in a patient of sub-saharan African origin. Radiol Case Rep. 2022;17(3) https://doi.org/10.1016/j.radcr.2021.11.046.
117. Mohamed C, Najib K, Essaadouni L. Radiological findings in Behçet disease. Pan Afr Med J. 2015:20. https://doi.org/10.11604/pamj.2015.20.51.5928.
118. Varga Z, Flammer AJ, Steiger P, Haberecker M, Andermatt R, Zinkernagel AS, et al. Endothelial cell infection and endotheliitis in COVID-19. Lancet. 2020;395(10234) https://doi.org/10.1016/S0140-6736(20)30937-5.
119. de Boysson H, Parienti JJ, Mawet J, Arquizan C, Boulouis G, Burcin C, Naggara O, Zuber M, et al. Primary angiitis of the CNS and reversible cerebral vasoconstriction syndrome: a comparative study. Neurology. 2018;91(16) https://doi.org/10.1212/WNL.0000000000006367.
120. Birnbaum J, Hellmann DB. Primary Angiitis of the central nervous system. Arch Neurol. 2009;66(6) https://doi.org/10.1001/archneurol.2009.76.
121. Sundaram S, Kumar PN, Sharma DP, Kesavadas C, Sreedharan SE, Prasad BA, et al. High-resolution Vessel Wall imaging in primary Angiitis of central nervous system. Ann Indian Acad Neurol. 2021;24(4) https://doi.org/10.4103/aian.AIAN_106_21.
122. Guo A, Zhang Z, Dong GH, Su L, Gao C, Zhang M, Shi X, Wang H, Zhang X, Lu DH, Fu Y, et al. Cortical microhemorrhage presentation of small vessel primary Angiitis of the central nervous system. Ann Neurol. 2024;96(1) https://doi.org/10.1002/ana.26940.
123. Sharma SK, Soneja M. HIV & immune reconstitution inflammatory syndrome (IRIS). Indian J Med Res. 2011;134(6) https://doi.org/10.4103/0971-5916.92632.
124. Sereti I, et al. Prospective international study of incidence and predictors of immune reconstitution inflammatory syndrome and death in people living with human immunodeficiency virus and severe lymphopenia. Clin Infect Dis. 2020;71(3) https://doi.org/10.1093/cid/ciz877.

125. Rose JJ, Wang L, Xu Q, McTiernan CF, Shiva S, Tejero J, et al. Carbon monoxide poisoning: pathogenesis, management, and future directions of therapy. Am J Respir Crit Care Med. 2017;195(5) https://doi.org/10.1164/rccm.201606-1275CI.
126. Beppu T. The role of MR imaging in assessment of brain damage from carbon monoxide poisoning: a review of the literature. Am J Neuroradiol. 2014;35(4) https://doi.org/10.3174/ajnr.A3489.
127. Leth PM, Gregersen M. Ethylene glycol poisoning. Forensic Sci Int. 2005;155(2–3) https://doi.org/10.1016/j.forsciint.2004.11.012.
128. Morgan BW, Ford MD, Follmer R. Ethylene glycol ingestion resulting in brainstem and midbrain dysfunction. J Toxicol Clin Toxicol. 2000;38(4) https://doi.org/10.1081/CLT-100100956.
129. Malhotra A, Mongelluzzo G, Wu X, Durand D, Kalra VB, LeSar B, et al. Ethylene glycol toxicity. Clin Neuroradiol. 2016;27:1. https://doi.org/10.1007/s00062-016-0525-0.
130. Manzano JMM, Elkholy KO, Lo KB. Brain MRI abnormalities in acute ethylene glycol poisoning: a case report. Toxicol Commun. 2022;6(1) https://doi.org/10.1080/24734306.2022.2087019.
131. Alrashed M, Aldeghaither NS, Almutairi SY, Almutairi M, Alghamdi A, Alqahtani T, et al. The perils of methanol exposure: insights into toxicity and clinical management. Toxics. 2024;12(12) https://doi.org/10.3390/toxics12120924.
132. Taheri MS, Moghaddam HH, Moharamzad Y, Dadgari S, Nahvi V. The value of brain CT findings in acute methanol toxicity. Eur J Radiol. 2010;73(2) https://doi.org/10.1016/j.ejrad.2008.11.006.
133. Cauter SV, Severino M, Ammendola R, Berkel BV, Vavro H, Lvd H, et al. Bilateral lesions of the basal ganglia and thalami (central grey matter) – pictorial review. Neuroradiology. 2020;62(12) https://doi.org/10.1007/s00234-020-02511-y.
134. Olmos-Jiménez R, Espuny-Miró A, Cárceles-Rodríguez C, Díaz-Carrasco MS. Practical aspects of the use of intrathecal chemotherapy. Farm Hosp. 2017;41(1) https://doi.org/10.7399/fh.2017.41.1.10616.
135. Panicker VV, Radhakrishnan SE, Kuruttukulam GV, Bose JA, Favas T. Methotrexate-induced leukoencephalopathy as a clinical and radiological mimicker of acute ischemic stroke leading to thrombolysis. Cureus. 2024;16(1) https://doi.org/10.7759/cureus.51542.
136. Inaba H, Khan RB, Laningham FH, Crews KR, Pui C-H, Daw NC. Clinical and radiological characteristics of methotrexate-induced acute encephalopathy in pediatric patients with cancer. Ann Oncol. 2008;19(1) https://doi.org/10.1093/annonc/mdm466.
137. Yoshimi M, Abe S. Diffusion-weighted imaging characteristics of methotrexate-induced acute encephalopathy. Clin Case Reports. 2018;6(3) https://doi.org/10.1002/ccr3.1376.
138. Apiraksattayakul N, Jitprapaikulsan J, Sanpakit K, Kumutpongpanich T, Apiraksattayakul N, Jitprapaikulsan J, et al. Potential neurotoxicity associated with methotrexate. Sci Rep. 2024;14(1) https://doi.org/10.1038/s41598-024-69263-0.
139. Bompaire F, Lahutte M, Buffat S, Soussain C, Ardisson AE, Terziev R, et al. New insights in radiation-induced leukoencephalopathy: a prospective cross-sectional study. Support Care Cancer. 2018;26(12) https://doi.org/10.1007/s00520-018-4296-9.
140. Oka M, Terae S, Kobayashi R, Sawamura Y, Kudoh K, Tha KK, Yoshida M, Kaneda M, et al. MRI in methotrexate-related leukoencephalopathy: disseminated necrotising leukoencephalopathy in comparison with mild leukoencephalopathy. Neuroradiology. 2003;45(7) https://doi.org/10.1007/s00234-003-0983-3.
141. Kim JY, Kim ST, Nam D-H, Lee J-I, Park K, Kong D-S. Leukoencephalopathy and disseminated necrotizing leukoencephalopathy following intrathecal methotrexate chemotherapy and radiation therapy for central nerve system lymphoma or leukemia. J Korean Neurosurg Soc. 2011;50(4) https://doi.org/10.3340/jkns.2011.50.4.304.
142. Katsura M, Sato J, Akahane M, Furuta T, Mori H, Abe O. Recognizing radiation-induced changes in the central nervous system: where to look and what to look for. Radiographics. 2020;41(1) https://doi.org/10.1148/rg.2021200064.

143. Knudsen GH, Kermanian N, Kock-Jensen CH, Antulov R. Opioid-induced leukoencephalopathies: a report of two cases. Case Rep Neurol. 2022;14(1) https://doi.org/10.1159/000521410.
144. Dang BTD, Pisinski L, Krauthamer AV, Ramachandran S. Heroin-induced leukoencephalopathy: chasing the imaging findings. Radiol Case Rep. 2024;19(9) https://doi.org/10.1016/j.radcr.2024.05.085.
145. Keogh CF, Andrews GT, Spacey SD, Forkheim KE, Graeb DA, Keogh CF, et al. Neuroimaging features of heroin inhalation toxicity: "chasing the dragon". Am J Roentgenol. 2012;180(3) https://doi.org/10.2214/ajr.180.3.1800847.
146. García-Valdecasas-Campelo E, González-Reimers E, Santolaria-Fernández F, De La Vega-Prieto MJ, Milena-Abril A, Sánchez-Pérez MJ, et al. Brain atrophy in alcoholics: relationship with alcohol intake; liver disease; nutritional status, and inflammation. Alcohol Alcohol. 2007;42(6) https://doi.org/10.1093/alcalc/agm065.
147. Zuccoli G, Gallucci M, Capellades J, Regnicolo L, Tumiati B, Giadás TC, et al. Wernicke encephalopathy: MR findings at clinical presentation in twenty-six alcoholic and nonalcoholic patients. Am J Neuroradiol. 2007;28(7) https://doi.org/10.3174/ajnr.A0544.
148. Rovira A, Alonso J, Cordoba J. MR imaging findings in hepatic encephalopathy. AJNR Am J Neuroradiol. 2008;29(9):1612–21. https://doi.org/10.3174/ajnr.A1139.
149. Lim CG, Hahm MH, Lee HJ. Hepatic encephalopathy on magnetic resonance imaging and its uncertain differential diagnoses: a narrative review. J Yeungnam Med Sci. 2023;40(2) https://doi.org/10.12701/jyms.2022.00689.

Index

G. W. Hooper (ed.), *The Neuronomicon*, https://doi.org/10.1007/978-3-032-05385-5

GPSR Compliance

The European Union's (EU) General Product Safety Regulation (GPSR) is a set of rules that requires consumer products to be safe and our obligations to ensure this.

If you have any concerns about our products, you can contact us on ProductSafety@springernature.com

In case Publisher is established outside the EU, the EU authorized representative is:

Springer Nature Customer Service Center GmbH
Europaplatz 3
69115 Heidelberg, Germany

Batch number: 10105676

Printed by Printforce, the Netherlands